Energy Efficiency and Sustenance

Energy Efficiency and Sustenance

Edited by **Patricia Sachs**

CLANRYE
INTERNATIONAL

New Jersey

Published by Clanrye International,
55 Van Reypen Street,
Jersey City, NJ 07306, USA
www.clanryeinternational.com

Energy Efficiency and Sustenance
Edited by Patricia Sachs

International Standard Book Number: 978-1-63240-560-9 (Hardback)

Printed in the United States of America.

Contents

Preface

I am honored to present to you this unique book which encompasses the most up-to-date data in the field. I was extremely pleased to get this opportunity of editing the work of experts from across the globe. I have also written papers in this field and researched the various aspects revolving around the progress of the discipline. I have tried to unify my knowledge along with that of stalwarts from every corner of the world, to produce a text which not only benefits the readers but also facilitates the growth of the field.

The concept of energy efficiency revolves around reducing energy consumption without compromising with the actual facilities used. Generally, the conventional products are replaced with more sophisticated ones such as a normal filament bulb replaced with energy efficient LED light. The techniques employed under this discipline may result in energy savings as well as financial cost savings. Sustenance is maintaining these practices in the long run. The topics covered in this extensive book deal with the core subjects related to energy efficiency and sustenance. A number of latest researches have been included to keep the readers up-to-date with the global concepts in this area of study. This book, with its detailed analyses and data, will prove immensely beneficial to professionals and students involved in this area at various levels. It is also a beneficial read for engineers, environmentalists, ecologists and conservationists.

Finally, I would like to thank all the contributing authors for their valuable time and contributions. This book would not have been possible without their efforts. I would also like to thank my friends and family for their constant support.

Editor

1

Solar Energy Sustainability in Jordan

Ahmad Qasaimeh[1], Mohammad Qasaimeh[2], Zaydoun Abu-Salem[3], Mohammad Momani[4]

[1]Department of Civil Engineering, Jerash University, Jerash, Jordan
[2]Chemical Engineering Department, AlHuson University College, Al-Balqa Applied University, Salt, Jordan
[3]Department of Civil Engineering, Philadelphia University, Amman, Jordan
[4]Department of Electrical Engineering, Yarmouk University, Irbid, Jordan
Email: argg22@yahoo.com

Abstract

Jordan is a country with highly fossil fuel deficiency and thus other energy sources are needed to be explored. Solar energy in Jordan is highly recognized as a good source of energy and an excellent substitute to the fossil fuel. The solar energy in this article is obtained via data bases and modeling techniques for the specified place coordinate and angle of inclination. The angles of sun irradiations are different throughout the year; therefore solar energy needs to be magnified by optimizing the angle of inclination of solar cells. In this research, the optimized angles throughout the year are obtained to be in the range: 10° - 60°. Solar energy can serve the residential building, the findings of this research show that every 1 m² of the solar cell may contribute to about 60% - 70% of customer needs of electricity throughout the year. The application of solar energy concept in the design of building will play an important role in energy sustainability.

Keywords

Solar Energy, Inclination Angle, Sustainable Buildings

1. Introduction

During the last two decades, the increasing energy demand has brought challenges to Jordan due to country's limited resources. Hence, solar energy applications have more attention to substitute the depletion of the fossil fuel that causes dramatic pollution to the environment. Jordan has established a strategic change and reform of its national economy and energy strategy [1]. Jordan has assisted programs utilizing solar energy. Assessment involved systematic monitoring of implementation of appropriate technologies, demonstrations, and pilot projects [2] [3].

Due to high and reliable solar irradiance in Jordan (5.5 kWh/m²·d), the usage for solar energy in Jordan has high potential for about 330 sunny days per year using solar panels [4]. Solar radiation also differs according to seasons, in winter the sun becomes lower in the sky and higher in summer because sun ray's angle changes due to the earth's tilt angle [5].

The distribution of total radiation on a horizontal surface over a day was examined by Liu and Jordan who showed that the ratio of hourly to daily radiation could be correlated with the local day length and angle which differs through the year [6]. Solar energy encounters many parameters that affect its cultivation during the year such as sunshine duration, relative humidity, temperature, and cloudiness. The solar panels inclination therefore should be dynamically changed during the seasons at certain place [7].

However solar radiation differs along the seasons. The results of Liu and Jordan were confirmed by Collares Pereira and Rabl [6] using a wider database for the average distribution of solar radiation associated with different coordinates of time and place. Saraf and Hamad [8] found the yearly optimum tilt angle in Basra, Iraq was higher than the latitude by about 8°. Gopinathan [9] showed the optimum tilt angle of oriented sloping plates is almost equal to place latitude.

Both Gopinathan [10] and Soulayman [11] showed the optimum tilt angle of oriented sloping plates is almost equal to place latitude.

However, researchers have different approaches for optimal angles for solar collectors in different places, because the radiation pattern changes from location to location and time to time [12] [13]. Thus, the aim of this research is to optimize the angle of inclination for Jordan during the year.

2. Methodology

The angles of sun irradiations are different throughout the year. Therefore, the solar panels should be dynamically inclined with different angles. This article spots the light on solar energy utilization depending on solar energy databases and modeling techniques. The information about solar energy, temperature, and electricity consumption was collected from several organizations namely: National Center for Research and Development, Ministry of Energy and Mineral Resources, and Jordan Meteorological Department. The sunshine hours in Jordan zone were taken from the time and date calendar. Solar Energy Modeling was performed using Meteonorm software Version 6 for modeling solar energy with inclination angle of panels for Jordan database.

3. Solar Energy for Buildings in Jordan

The major goal of this research is to explore that the collected solar energy can offset the electrical energy consumption in residential buildings (**Table 1**) [14]. The aim of the research can be achieved via many tracts. The first tract is to optimize solar panel angle of inclination throughout the year. The second tract is to design sufficient area in the residential buildings for panels to be installed. The third tract is to manage the energy in the building in the basis of building design and geometry, and daily wise management of energy.

Fossil fuel depletes and costly increases with the time, furthermore it causes environmental problems such as global warming. Therefore, solar cells must be oriented and distributed effectively depending on time in the year and depending on the building size. The savings of electricity can be enhanced by altering the daily time of wakeup and sleep. The daylight hours may help in utilizing the natural sunlight instead of electricity [15]. The implementation of Daylight Saving Time (DST) creates an additional hour of higher outdoor air temperature and solar radiation during the primary cooling times of the evening [16]. California Energy Commission [17] [18] conducted a simulation-based study to examine the effects of DST on statewide electricity consumption. Consequently, by concise management, collecting sun irradiation and fitting the daily man activities to sunshine will compensate large part of electricity for residential building.

The records about the solar energy in Jordan spots the light on the truth that Jordan is rich in sun irradiations as **Table 2** shows the data about sunshine hours and solar energy in different places in Jordan in the year [19]. The average of the data values shows the relative trend between solar energy and sunshine period as it is shown in **Figure 1**. The solar energy data accumulates an estimated average energy value on the yearly basis of about 2056 kWh/m^2 for Jordan.

In **Figure 1**, the information obtained about the solar energy can be represented via analytical model that incorporates the sunshine period as a major parameter. After calibration, this model can forecast the energy for specific place coordination and variable angle of inclination.

Table 1. The annual electricity consumption per capita in residential building in Jordan.

Equipment	Service (hours/day)		Operation	Consumption	Percent of Consumption
	Summer	Winter	(hours/year)	(kWh)	
Lighting	7	6	2340	514.8	21.28%
Refrigerator	16	8	4392	933.3	38.58%
TV	12	7	3372	288.3	11.92%
W. Machine	2	2	220	89.1	3.68%
Iron	1	1	130	117	4.83%
Fan	14	0	420	23.5	0.97%
Water Pump	1.5	1.5	135	10.8	0.45%
Freezer	10	2	2160	113.4	4.69%
Water Cooler	2	0	90	0.9	0.04%
Vacuum Cleaner	1	1	120	24	0.99%
Washing Dryer	0.5	0.5	180	5.4	0.22%
Hear Dryer	0.5	0.5	50	7.5	0.31%
Heater	1	1	60	2.7	0.11%
Geyser	1	5	480	192	7.94%
Air Condition	10	0	350	96.3	3.98%
			Total	2419	100%

4. Solar Energy Modeling throughout the Year

In this section, the solar energy in Jordan is being modeled for different angle of solar panel inclination. Meteonorm version 6 uses data bases about the specified place coordination and gives the estimated solar energy for different panel inclinations. It comprises physical and environmental parameters applicable for certain coordinate. The estimation of solar energy is being adjusted for different angle of orientation of solar panel due to sun movement during the months of the year. The modeling process is shown on **Table 3** that depicts the solar energy throughout the months in the year for different solar cell inclination.

The energy values in **Table 3** can be optimized to maximum solar energy for each month, for example for January the maximum energy is obtained when the panel is inclined to 60°. **Figure 2** shows the optimum solar energy with optimum solar cell inclination in each month in Jordan.

In this research, the focus is to optimize solar energy cultivation along the year. Hence, the angle of inclination is given the great attention. The data recorded about solar energy is given via panels of fixed inclination (**Table 2**). **Figure 3** compares between the energy recorded data for fixed panels and the dynamic estimated energy values for variable angle of inclination during the year.

The variable angle of inclination represents other important parameter of modeling in addition to the sunshine period. As the sun moves in orbital track, the sunshine period and the sun irradiations direction are variable along the year, which creates the seasons. In Jordan, the year is classified into four seasons: Summer, fall, winter, and spring. The summer season extends from May until August. The fall is characterized to be in September and October. Winter extends from November until February, and spring denotes to March and April. **Figure 4** represents each season and the solar energy accumulated with it within variable angle of inclination. For example, the figure shows that the solar energy gained in summer is as high as 1002 kWh/m^2.

As per for **Figure 4**, the optimized energy upon yearly basis is computed to be 2501 kWh/m^2 and this is also illustrated in **Table 4**. This value overcomes the value of general solar fixed cells (2056 kWh/m^2) that seen in **Figure 1**. By comparing the total energy value (2501 kWh/m^2—**Table 4**) and the value of electricity consumption

Table 2. Data of sun shine period (hours) and energy (kWh/m^2) during the report in 2007 in different stations in Jordan.

Sunshine	Amman	Irbid	Dhlail	Azraq
JAN	5.1	5.5	5.5	6
FEB	5.8	5.2	6.5	6.6
MAR	7.7	7.5	8.6	8.4
APR	9.4	9	9.7	9.3
MAY	9.8	9.6	9.9	9.6
JUN	11.8	11.3	11.7	11.1
JUL	12	11.8	12	11.2
AUG	11.7	12.2	11.6	11.1
SEPT	10	10	11	9.7
OCT	8.8	8.5	9	8.7
NOV	6.2	6.3	6.7	7
DEC	5	4.2	5.7	5.5
Energy	**Amman**	**Irbid**	**Dhlail**	**Azraq**
JAN	3.2	3.1	3.2	3.3
FEB	4.1	3.7	4.1	4.2
MAR	5.4	5.1	5.5	5.7
APR	6.8	6.4	6.8	6.8
MAY	7.1	6.9	7.5	7.5
JUN	8.3	7.7	8.4	8.4
JUL	8.1	7.6	8.3	8.1
AUG	7.4	7.5	7.7	7.3
SEPT	6	6	6.5	6.2
OCT	4.5	4.6	4.9	4.9
NOV	3.3	3.3	3.3	3.7
DEC	2.9	2.9	2.9	2.9

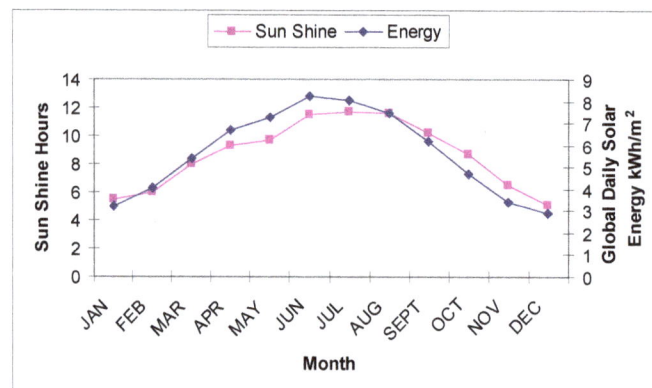

Figure 1. Average sunshine hours and global daily solar energy recorded in Jordan.

Table 3. The estimated solar energy kWh/m² during months of the year for different solar cell inclination.

	Solar Energy kWh/m²					
	10°	20°	30°	40°	50°	60°
JAN	119	136	149	158	163	164
FEB	123	135	143	148	149	147
MAR	175	184	189	189	185	176
APR	204	207	205	198	186	170
MAY	241	236	225	209	189	163
JUN	254	244	228	207	182	153
JUL	262	253	239	219	194	164
AUG	245	245	238	226	208	185
SEP	216	226	229	227	218	204
OCT	185	203	215	221	222	216
NOV	136	154	169	178	183	183
DEC	111	127	140	149	155	156

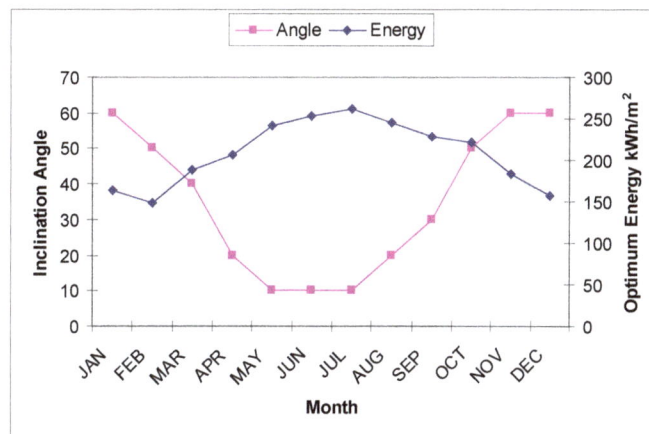

Figure 2. Optimum solar energy with optimum solar cell inclination in each month in Jordan.

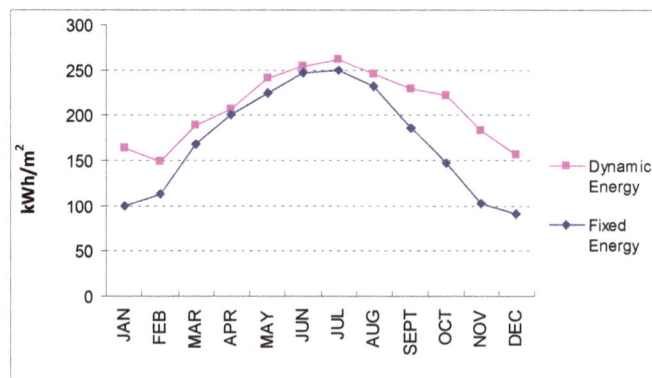

Figure 3. The comparison between energy gathered by panels of dynamic angle of inclination and panels of fixed angle of inclination.

Table 4. The optimal solar energy cultivated during months in Jordan.

MONTH	Inclination Angle	Optimal Solar Energy (kWh/m^2)
JAN	60	164
FEB	50	149
MAR	40	189
APR	20	207
MAY	10	241
JUN	10	254
JUL	10	262
AUG	20	245
SEP	30	229
OCT	50	222
NOV	60	183
DEC	60	156
Total Energy		**2501**

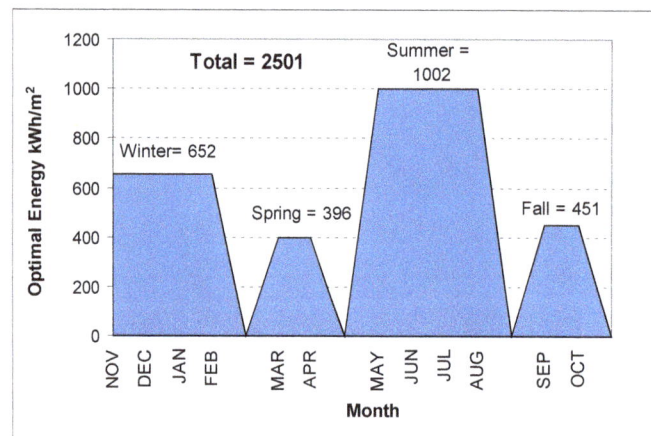

Figure 4. The optimal solar energy cultivated in each season in Jordan.

for residential building per year for single customer (2419 kWh/capita—**Table 1**); it is worthy to take in consideration that if the solar energy is efficiently converted to electricity then every 1 m^2 of the solar cell will contribute to about 60% - 70% of customer needs of energy in the year. Consequently, the area occupied by solar panels is another important parameter in energy estimation. Hence, the idea of utilizing solar energy in buildings is an important scenario for sustainability.

5. Conclusions

Jordan is a place that it is recognized of plentiful high solar radiation. The solar energy depends on the sun travel along the years, and thus the sunshine period and sun irradiations direction are variable during the year. Solar energy cultivation is optimized considering dynamic variation of the angle of solar cell inclination through the year.

In this research, it's shown that solar energy can serve the residential building consumption of electricity as every 1 m^2 of the solar cell may contribute to about 60% - 70% of customer needs of energy in the year. In addition, combining solar energy concept in the design of building will lower the cost of energy use and will play an important role in sustainable development of buildings.

References

[1] Abdelkader, M.R., Al-Salaymeh, A., Al-Hamamre, Z. and Sharaf, F. (2010) A Comparative Analysis of the Performance of Monocrystalline and Multiycrystalline PV Cells in Semi Arid Climate Conditions: The Case of Jordan. *Jordan Journal of Mechanical and Industrial Engineering*, **4**, 543-552.

[2] Blumenberg, J., Bentenrieder, M. and Kerschensteiner, H. (1997) Introducing Advanced Testing Methods for Domestic Hot Water Storage Tanks in Jordan. *Renewable Energy*, **10**, 207-211.

[3] Badran, A. (2001) A Study in Industrial Applications of Solar Energy and the Range of Its Utilization in Jordan. *Renewable Energy*, **24**, 485-490. http://dx.doi.org/10.1016/S0960-1481(01)00032-5

[4] Shariah, A., Al-Akhras, M.-A. and Al-Omari, I.A. (2002) Optimizing the Tilt Angle of Solar Collectors. *Renewable Energy*, **26**, 587-598. http://dx.doi.org/10.1016/S0960-1481(01)00106-9

[5] Jibril, Z. (1991) Estimation of Solar Radiation over Jordan Predicted Tables. *Renewable Energy*, **1**, 277-291. http://dx.doi.org/10.1016/0960-1481(91)90087-6

[6] Collares-Pereira, M. and Rabl, A. (1979) The Average Distribution of Solar Radiation-Correlations between Diffuse and Hemispherical and between Daily and Hourly Insolation Values. *Solar Energy*, **22**, 155-164. http://dx.doi.org/10.1016/0038-092X(79)90100-2

[7] Singh, H.N. and Tiwari, G.N. (2005) Evaluation of Cloudiness, Haziness Factor for Composite Climate. *Energy*, **30**, 1589-1601.

[8] Saraf, G.R. and Hamad, F.A.W. (1988) Optimum Tilt Angle for a Flat Plate Solar Collector. *Energy Conversion and Management*, **28**, 185-191. http://dx.doi.org/10.1016/0196-8904(88)90044-1

[9] Gopinathan, K.K. (1988) A General Formula for Computing the Coefficients of the Correlation Connecting Global Solar Radiation to Sunshine Duration. *Solar Energy*, **41**, 499-502. http://dx.doi.org/10.1016/0038-092X(88)90052-7

[10] Gopinathan, K.K. (1991) Solar Radiation on Variously Oriented Sloping Surfaces. *Solar Energy*, **47**, 173-179. http://dx.doi.org/10.1016/0038-092X(91)90076-9

[11] Soulayman, S.Sh. (1991) On the Optimum Tilt of Solar Absorber Plates. *Renewable Energy*, **1**, 551-554. http://dx.doi.org/10.1016/0960-1481(91)90070-6

[12] Gunerhan, H. and Hepbasli, A. (2007) Determination of the Optimum Tilt Angle of Solar Collectors for Building Applications. *Building and Environment*, **42**, 779-783. http://dx.doi.org/10.1016/j.buildenv.2005.09.012

[13] Yakup, M.Ab.H.M. and Malik, A.Q. (2001) Optimum Tilt Angle and Orientation for Solar Collector in Brunei Darussalam. *Renewable Energy*, **24**, 223-234. http://dx.doi.org/10.1016/S0960-1481(00)00168-3

[14] Ministry of Energy and Mineral Resources, MEMR (1996) *Analytical Study Report*. Jordan.

[15] Momani, M.A., Yatim, B. and Mohd Ali, M.A. (2009) The Impact of the Daylight Saving Time on Electricity Consumption—A Case Study from Jordan. *Energy Policy*, **37**, 2042-2051.

[16] Shimoda, Y., Asahi, T., Taniguchi, A. and Mizuno, M. (2007) Evaluation of City-Scale Impact of Residential Energy Conservation Measures Using the Detailed End-Use Simulation Model. *Energy*, **32**, 1617-1633. http://dx.doi.org/10.1016/j.energy.2007.01.007

[17] Kandel, A. and Metz, D. (2001) The Effects of Daylight Saving Time on California Electricity Use. California Energy Commission (CEC), Sacramento.

[18] Kandel, A. (2001) Electricity Savings from Early Daylight Saving Time. California Energy Commission (CEC), Sacramento.

[19] Jordan Meteorological Department (JMD) (2007) Jordan Annual Climate Bulletin. Jordan Meteorological Department, Amman.

2

Parametric Study on Phase Change Material Based Thermal Energy Storage System

Kondakkagari Dharma Reddy*, Pathi Venkataramaiah, Tupakula Reddy Lokesh

Department of Mechanical Engineering, S. V. University, Tirupati, India
Email: *kdharmareddy@gmail.com

Abstract

The usage of phase change materials (PCM) to store the heat in the form of latent heat is increased, because large quantity of thermal energy is stored in smaller volumes. In the present experimental investigation, sodium thiosulphate pentahydrate is employed as phase change material and it is stored in stainless steel capsules. These capsules are kept in fabricated tank and hot water is supplied into it. The experimental design is prepared by considering the parameters: flow rate, heat transfer fluid inlet temperature and PCM capsule shape. Experiments are conducted according to the experimental design and responses are recorded. The effect of selected parameters on TES using PCM is studied by analyzing experimental data. The experimental data are also analyzed using Fuzzy Logic to find the optimal values of flow rate, heat transfer fluid inlet temperature and PCM capsule shapes. The present work utilizes Fuzzy Logic to find the optimal parameters for designing the effective Thermal Energy Storage System (TES).

Keywords

Phase Change Material (PCM), Thermal Energy Storage System, Fuzzy Logic

1. Introduction

The continuous increase in the level of greenhouse emissions and the rise in fuel prices are the main driving forces behind efforts to more effective utilization of various sources of renewable energy. Energy storage units can be used to reduce energy consumption by using available waste heat or alternate energy sources.

This also leads to saving of primary fuels and makes the system more cost effective by reducing the wastage of energy. The energy storage can also even out the mismatch between energy supply and consumption and thereby helps in saving capital costs. Thermal energy storage (TES) is one of the key technologies for energy conservation and is used to assist in the effective utilization of thermal energy in a wide number of applications.

---

*Corresponding author.

2. Literature Review and Objective

The behavior of a packed bed latent heat thermal energy storage system is analyzed. The packed bed utilizes the spherical capsules filled with paraffin wax as phase change material (PCM) usable with solar water heating system. The equations are numerically solved, and the results obtained are used for the thermal performance analysis of both charging and discharging process. The effect of inlet heat transfer fluid temperature (Steffan number), mass flow rate and phase change temperature on the thermal performance of capsules of different radii have been investigated [1]. The application of Taguchi's robust design coupled with fuzzy based desirability function approach for optimizing multiple bead geometry parameters of submerged arc weldment and Fuzzy inference system has been adapted to avoid uncertainly, imprecision and vagueness in experimentation as well as in data analysis by traditional Taguchi based optimization approach [2].

   The solar water heating system incorporating with phase change materials (PCMs) had been investigated [3]. An attempt was made to investigate and analyze the available thermal energy storage systems Incorporating PCMs for use in different applications. [4] investigated and analyzed the thermal energy storage extracted from solar heater and use for domestic purpose and Phase change materials as paraffin wax and sodium acetate try hydrate are used for storing thermal energy in two different insulation tanks [5]. A spherical container, filled with spheres containing either paraffin wax or stearic acid as Phase Change Materials (PCM) and water occupying the space left between the spheres had been studied and ANSYS is used for modeling the PCM inside the spheres with simple 2D. [6] studied the feasibility of storing solar energy using Phase Change Materials (PCMs) and utilizing this energy to heat water for domestic purposes during night time and this ensures that hot water is available throughout the day. The storage unit utilizes small cylinders, made of aluminium, filled with paraffin wax as the heat storage medium. [7] implemented and tested thermal energy storage (TES) system using different phase change materials (PCM) for solar cooling and refrigeration applications. A high temperature pilot plant is able to test different types of TES systems and uses synthetic thermal oil as heat transfer fluid (HTF). Two different PCM were selected after a deep study of the requirements of a real solar cooling plant and the available materials in the market, finally d-mannitol with phase change temperature of 167°C and hydroquinone which has a melting temperature of 172.2°C were used. [8] studied Fuzzy Logic integrated with the Taguchi method is used to optimize Wire Electro Discharge Machining (WEDM) process with multiple quality characteristics. The application of the Taguchi method with Fuzzy Logic was discussed to optimize the machining parameters for Wire electrical discharge machining of Inconel 825 with multiple characteristics. A multi-response performance index (MRPI) was used for optimization. The machining parameters viz., pulse on time, pulse off time, corner servo voltage, flushing pressure, wire feed, wire tensiospark gap voltage, servo feed were optimized with consideration of multiple performance characteristics. [9] combined Fuzzy Logic with the grey relational analysis, the designed algorithm is transformed into optimization of a single and simple grey-fuzzy reasoning grade rather than multiple performance characteristics. The Taguchi method is also adopted to search for an optimal combination of cutting parameters [10]. According to [11], solar water heater is getting popularity since they are relatively inexpensive and simple to fabricate and maintain. Studied building integrated solar energy collection system into the building shell and mechanical systems may reduce the cost of the solar energy systems and improve the efficiency of the collection. [12] developed a surface roughness prediction model using Fuzzy Logic for end milling of Al-SiCp metal matrix composite using carbide end mill cutter. The surface roughness is modeled as a function of spindle speed (N), feed rate (f), depth of cut (d) and the SiCp percentage (S). The predicted values surface roughness is compared with experimental result. It is observed that surface roughness is most influenced by feed rate, spindle speed and SiC percentage. Depth of cut has least influence. [13] described the importance of Latent heat storage, according to his review Paraffin waxes are cheap and have moderate thermal energy storage density but low thermal conductivity and, hence, require a large surface area. Hydrated salts have a larger energy storage density and a higher thermal conductivity. [14] Described the basic requirements of phase change materials with respective to their most important properties like latent heat, density etc, advantages, and disadvantages. [15] studied three modes of thermal energy storage (TES), and these are sensible heat storage (SHS), latent heat storage (LHS), and bond energy storage (BES). The SHS refers to the energy systems that store thermal energy without phase change. Heating of a material that undergoes a phase change (usually melting) is called the LHS. In the LHS, the storage operates isothermally at the phase change of the material. Lastly, comparison of storage system types is also presented. [16] Optimized the drilling characteristics for Al/SiCp composites using Fuzzy Logic and genetic algorithms (GA) and thedrilling characteristics were drill wear, specific energy and surface roughness [17]. Deals with the melting phenomena of Phase Change

Material (PCM) need to be understood for the design of thermal storage systems. The constrained and unconstrained melting of PCM inside a spherical capsule using paraffin wax (PW) is investigated. The experiments are carried out with different HTF temperatures of 62°C, 70°C, 75°C and 80°C. [18] studied the feasibility of storing solar energy using phase change materials (PCMs) stored in small cylinders and utilizing this energy to heat water for domestic applications during night time. [19] discussed the physical reasons for phase changing property of a phase change material (PCMs) and some important applications of PCMs [20]. After a rigorous study of their properties like melting temperature, heat of fusion, thermal conductivity and density, a concluding list of nine promising phase changing materials appropriate for thermal energy storage is prepared [21].

The literature review reveals that no researcher was used the Fuzzy Logic technique for optimizing the thermal energy storage system parameters using different shapes of the capsules in which PCM is stored. In this research work, an attempt was made to determine the optimum levels of the parameters of thermal energy storage system using Fuzzy Logic.

3. Experimental Setup

A schematic diagram of the experimental set-up is shown in **Figure 1**. This consists of an insulated cylindrical TES tank, which contains PCM capsules (cylindrical, spherical, and square capsules), flow meter and water storage tank. The stainless steel TES tank has a capacity of 10 liters. The storage tank is insulated with glass wool of 50 mm thick. The PCM capsules of different shapes are uniformly packed in the storage tank. The $Na_2S_2O_3 \cdot 5H_2O$ is used as PCM that has a melting temperature of 48°C and latent heat of fusion of 210 kJ/kg. Water is used as both SHS material and HTF.

A flow meter with an accuracy of ±2% is used to measure the flow rate of HTF and the flow rate is changed by different tap openings. The TES tank is incorporated with digital thermometers with an accuracy of ±1°C is placed above the TES tank to measure the temperatures of HTF and PCM stored inside the capsules. An electric water heater is used to maintain the constant temperature in the water storage tank. The thermo-physical properties of PCM are given in **Table 1**.

1. Electric heater; 2. Constant temperature bath (water storage tank); 3 & 8 Flow control valves; 4. Flow meter; 5. Distributer; 6. TES Tank; 7. PCM capsules; 9. Outlet tank; 10. Digital thermometer.

3.1. Experimental Trail

The experiments are carried out on the basis Taguchi design obtained from Minitab by considering the factors flow rate of HTF, HTF inlet temperature and PCM capsule shapes at different levels as shown in **Table 2**. And the measured responses *i.e.*, charging time and discharging time is shown in **Table 3**.

Table 1. Thermo-physical properties of PCM.

	Melting temperature (°C)	Latent heat of fusion (kJ/kg)	Density (kg/m³)		Specific heat (J/kg·°C)	
			Solid	Liquid	Solid	Liquid
Sodium thiosulfate pentahydrate ($Na_2S_2O_3 \cdot 5H_2O$)	48	210	1750	1670	2.38	1.46

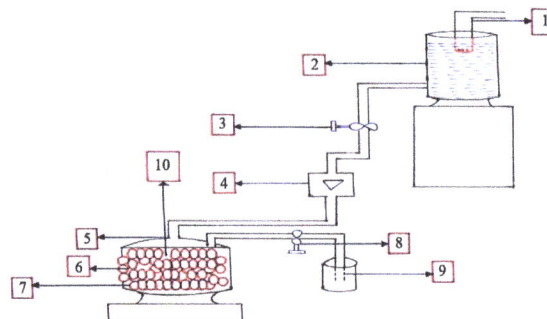

Figure 1. Schematic diagram of experimental setup.

Table 2. Factors and their levels.

Sl. No		Factors		
	Levels	Flow rate (lit/min)	HTF inlet temperature (°C)	PCM capsule shape
1	**1**	2	56	Cylindrical
2	2	4	58	Square
3	3	-	60	Sphere

Table 3. Experimental layouts and measured response values.

Experimental run	Inputs			Responses	
	Flow rate (lit/min)	HTF inlet temp (T_{HTF}) (°C)	PCM capsule shape	Charging time (min)	Discharging time (min)
1	2	60	1	10	140
2	2	60	2	32	140
3	2	60	3	18	100
4	2	58	1	14	100
5	2	58	2	34	120
6	2	58	3	18	80
7	2	56	1	16	100
8	2	56	2	36	100
9	2	56	3	18	80
10	4	60	1	08	140
11	4	60	2	28	140
12	4	60	3	12	100
13	4	58	1	11	100
14	4	58	2	32	120
15	4	58	3	18	80
16	4	56	1	16	100
17	4	56	2	32	100
18	4	56	3	18	80

3.2. Charging Process

During the charging process (storing of heat energy), the HTF is circulated through the TES tank continuously. The HTF exchanges its energy to PCM capsules and at the beginning of the charging process, the temperature of the PCM (T_{PCM}) inside the packed bed capsules is 32°C, which is lower than the melting temperature. Initially the energy is stored inside the capsules as sensible heat until the PCM reaches its melting temperature. As the charging process proceeds, energy storage is achieved by melting the PCM at a constant temperature. Finally, the PCM becomes superheated. The energy is then stored as sensible heat in liquid PCM. Temperatures of the PCM and HTF are recorded at an interval of 2 minutes. The charging process is continued until the PCM temperature comes in equilibrium with the temperature of HTF in the TES tank. The key experimental parameters HTF inlet temperature, flow rate of HTF and PCM capsule shapes are studied by considering the charging time.

3.3. Discharge Process

The discharging process started after the completion of charging process. Batch wise discharging experiments are carried out. In this method, 2 liters of hot water is discharged from the thermal energy storage tank and the same quantity of cold water at 32°C is fed into TES tank in each batch. The average temperature of the collected discharge water in the bucket is measured using a digital thermometer. The time difference between the consequent discharges is 20 min. The batch wise withdrawing of hot water is continued till the temperature of the outlet water reaches 32°C. A comparative study is also made between the conventional SHS system and combined storage system (SHS + LHS).

4. Results and Discussions

4.1. Charging Process

4.1.1. Effect of Flow Rate on Different PCM Capsules

Figures 2-4 illustrate the effect of varying the mass flow rate of HTF (2 and 4 lit/min) during the charging of the storage tank. Increase in mass flow rate has a large influence on the phase transition process of PCM. As the flow rate increases, the time required to complete charging becomes smaller.

When the flow rate increases from two lit/min to four lit/min the charging time is reduced by 20%, 12.5% and 25% for cylindrical, spherical and square PCM capsules respectively. Hence, mass flow rate has a significant effect on the charging process of thermal energy storage tank.

4.1.2. Effect of HTF Inlet Temperature on Different PCM Capsules

Figures 5-7 indicate the effects of HTF inlet temperature on different capsule shapes when the HTF inlet

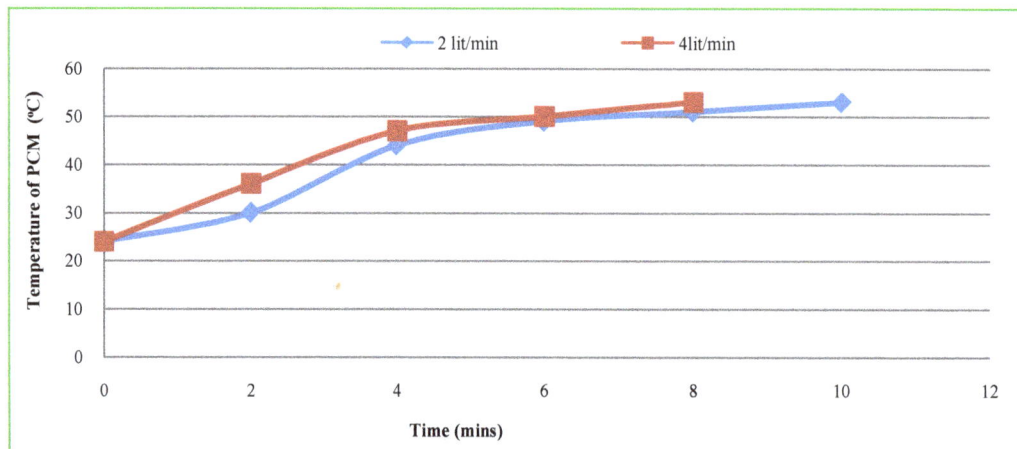

Figure 2. Effect of flow rates on cylindrical PCM capsules.

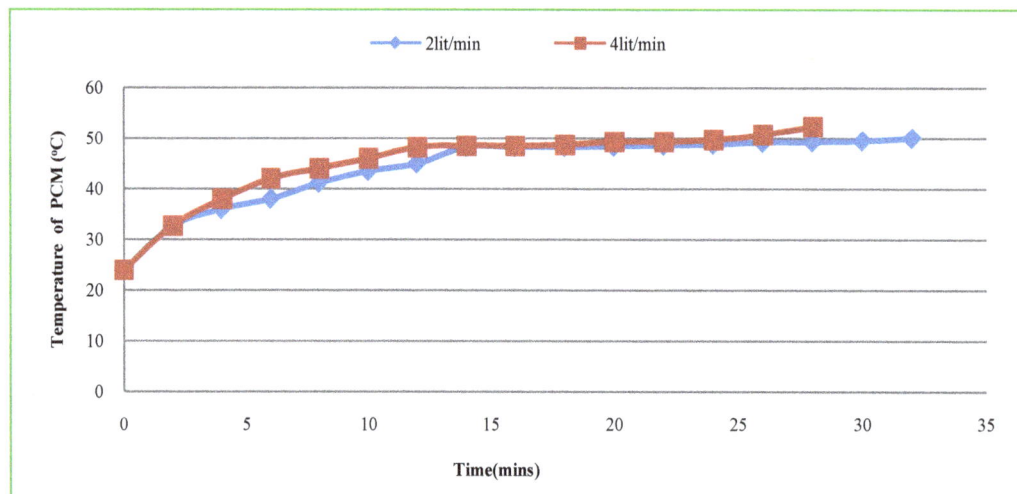

Figure 3. Effect of flow rates on spherical PCM capsules.

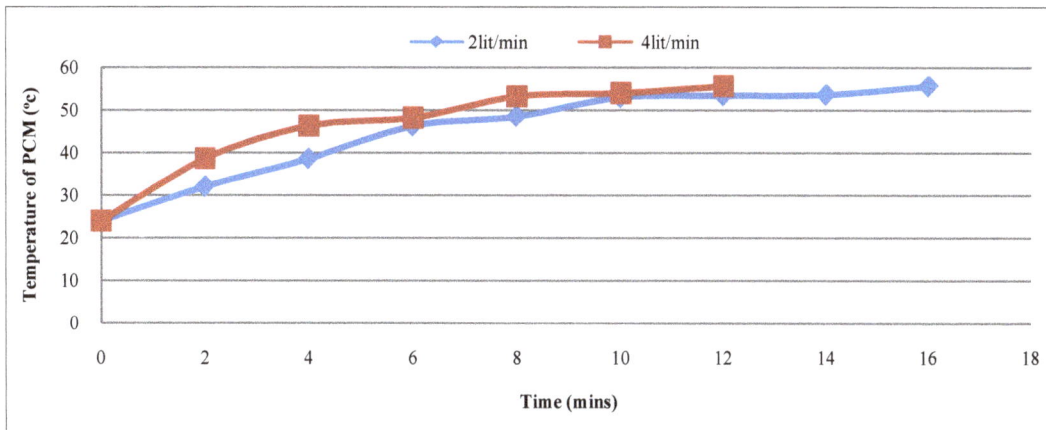

Figure 4. Effect of flow rates on square PCM capsules.

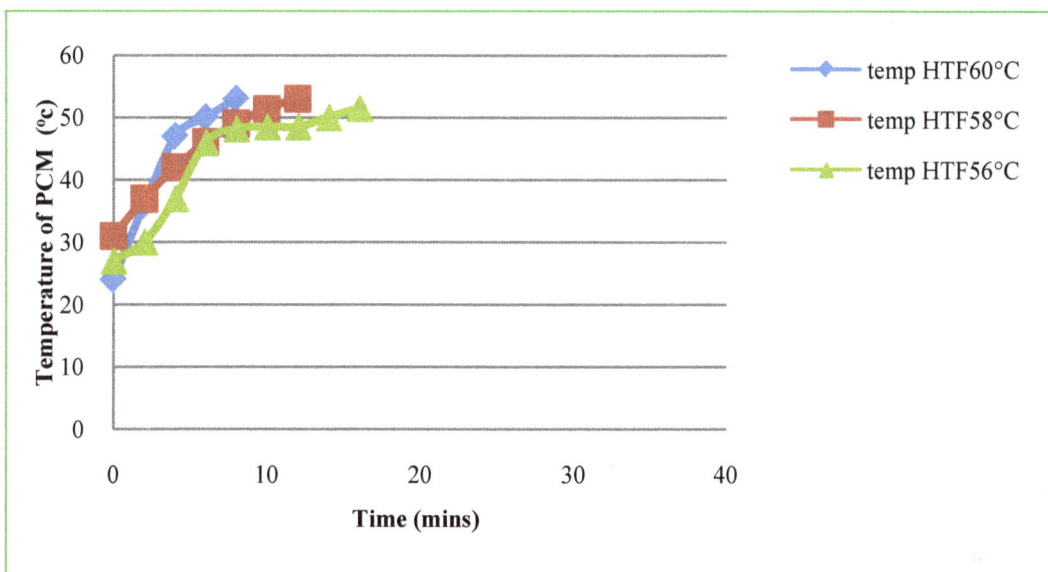

Figure 5. Effect of HTF inlet temperatures for cylindrical capsules.

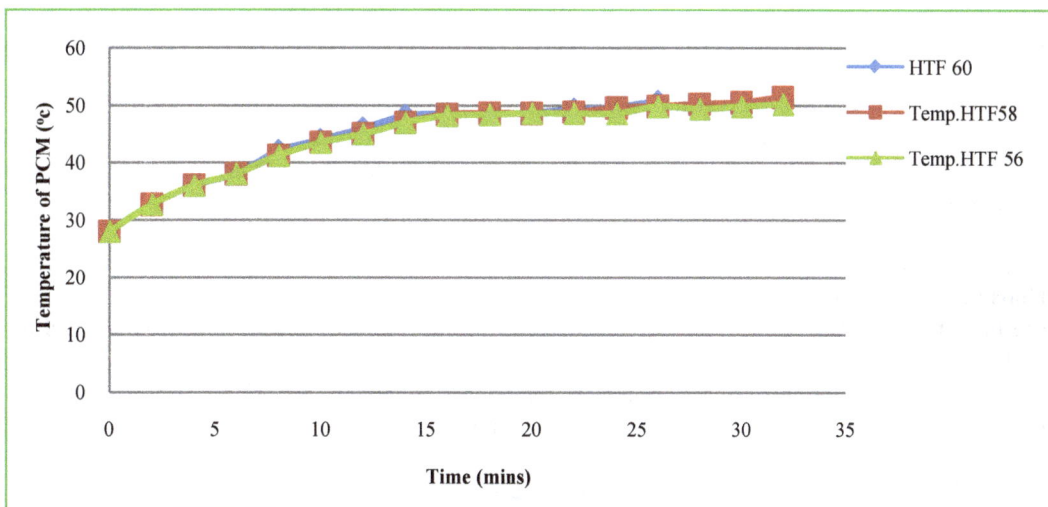

Figure 6. Effect of HTF inlet temperature for spherical capsules.

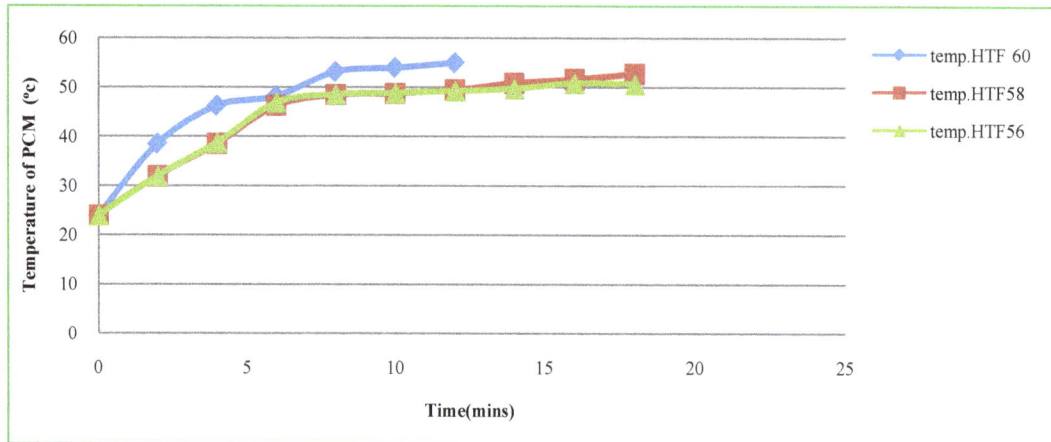

Figure 7. Effect of HTF inlet temperatures for square capsules.

temperature increases from 56°C to 60°C. The charging process is reduced by 50%, 12.5% and 33.33% when cylindrical capsules, spherical capsules and square capsules are employed. These shows the charging time is strongly affected by the inlet temperature of heat transfer fluid. It is concluded that higher the inlet heat transfer fluid temperature, the shorter the time interval to complete charging process.

4.1.3. Effect of Capsule Shape
The effect of capsule shape is studied by considering the best flow rate and best heat transfer fluid inlet temperatures. The graphs shown below are drawn by considering HTF inlet temperature at 60°C and mass flow rate of HTF at 4 lit/min.

From **Figure 8**, it is observed that, the cylindrical capsules reduces the charging time by 71.43%, 33.33% when compared to spherical and square capsules respectively. And also surface area to volume ratio for cylindrical capsule (25 mm dia) is 180 m^{-1} and for spherical capsule (50 mm dia) is 120 m^{-1}. And for square capsule is 33.33 m^{-1}. Hence, it is concluded that capsule having high surface area to volume ratio absorbs more thermal energy in a given time one leading to lesser charging time, as it presents more surface area for a given volume. Hence cylindrical capsule is selected as the best capsule shape in which PCM is to be stored.

4.1.4. Discharging Process
In this process, a certain amount of hot water (say 2 lit) is withdrawn from the TES tank and same amount of water is fed into the TES tank and the temperature of withdrawn water is noted. The water is withdrawn from the TES tank for every 20 minutes. This entire process is called batch process. The same procedure is repeated for different capsule shapes. The discharging process is continued until the temperature of withdrawn water (T_W) reached to 32°C.

From **Figure 9**, it is observed that the energy retrieval time or discharging time is more for cylindrical capsules and spherical capsules when compared to square capsules and also same for cylindrical and spherical capsules. But cylindrical capsules require less charging time and have high surface area to volume ratio. Hence cylindrical capsule (25 mm dia.) is considered to be the best capsule shape to store phase change material.

4.1.5. Comparison with SHS System
Discharging Process in SHS System
The results obtained from SHS system are compared with the combined system and are tabulated in **Table 4**. From **Table 4**, it is observed that at a constant charging temperature of 60°C, the thermal energy retrieved is 1484.68 KJ from the combined system where as the energy retrieved 1406.83 KJ is only from SHS system, hence a combined system is more efficient and recommended.

4.2. Selection of Parameters Using Fuzzy Logic
The experimental results are analyzed using Fuzzy Logic to select the optimum parameters. Smaller the better

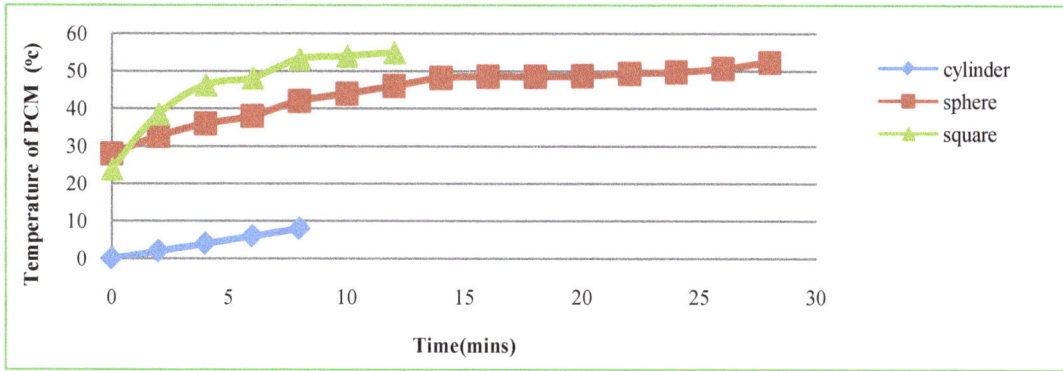

Figure 8. Effect of charging time on different capsule shapes.

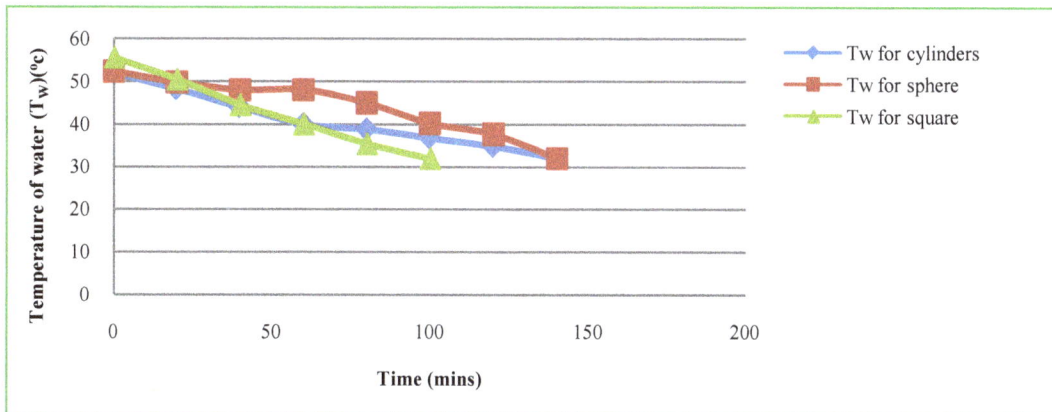

Figure 9. Effect of discharging time on different capsule shapes.

Table 4. Comparison of (SHS + LHS) and SHS.

Status of description	TES (SHS + LHS)	SHS
Total volume of the TES tank (lit)	10	10
Total hot water withdrawn	16	12
Energy stored (KJ)		
Energy stored in the sodium thiosulfate pentahydrate	57.12	
Total energy stored	1660.20	1172.08
Heat recovered from the system in the form of Hot water during discharge	1484.68	1406.83
Energy retrieval time (min)	140	100

formula is used to find S/N ratios of the experimental values of charging time. To find S/N ratios of the experimental results of discharging time, the larger the better formula is used and the values are tabulated in **Table 5**.

The structure built for this study is a two input-one-output Fuzzy Logic unit as shown in **Figure 10(a)**. The input variables of the Fuzzy Logic system in this study are the S/N ratios of responses charging time and discharging time. They are converted into linguistic fuzzy subsets using membership functions of a triangle form, as shown in **Figure 10(b)**, and are uniformly assigned into three fuzzy subsets—low (L), medium (M) and high (H) grade. The output variable of this analysis is the comprehensive output measure (COM) and the comprehensive output measure (COM) is generated using MATALAB, which is shown in **Table 6**. Membership functions used for this work are of a triangle form, as shown in **Figure 10(b)**. Unlike the input variables, the output variable is assigned into relatively nine subsets *i.e.*, very very low (VVL), very low (VL), small (S), medium low

Table 5. S/N ratios of experimental values.

Experimental run	Flow rate	HTF inlet temp (T_{HTF})	Capsule shape	Charging time	Discharging time	S/NRA1	S/NRA2
1	2	60	1	10	140	−20	42.92256
2	2	60	2	32	140	−30.103	42.92256
3	2	60	3	18	100	−25.1055	40
4	2	58	1	14	100	−22.9226	40
5	2	58	2	34	120	−31.1261	40
6	2	58	3	18	80	−25.1055	38.0618
7	2	56	1	16	100	−24.0824	40
8	2	56	2	36	100	−31.1261	40
9	2	56	3	18	80	−25.1055	38.0618
10	4	60	1	08	140	−18.0618	42.92256
11	4	60	2	28	140	−28.9432	42.92256
12	4	60	3	12	100	−21.5836	40
13	4	58	1	11	100	−21.5836	40
14	4	58	2	32	120	−30.103	40
15	4	58	3	18	80	−25.1055	38.0618
16	4	56	1	16	100	−22.9226	40
17	4	56	2	32	100	−30.103	40
18	4	56	3	18	80	−25.1055	38.0618

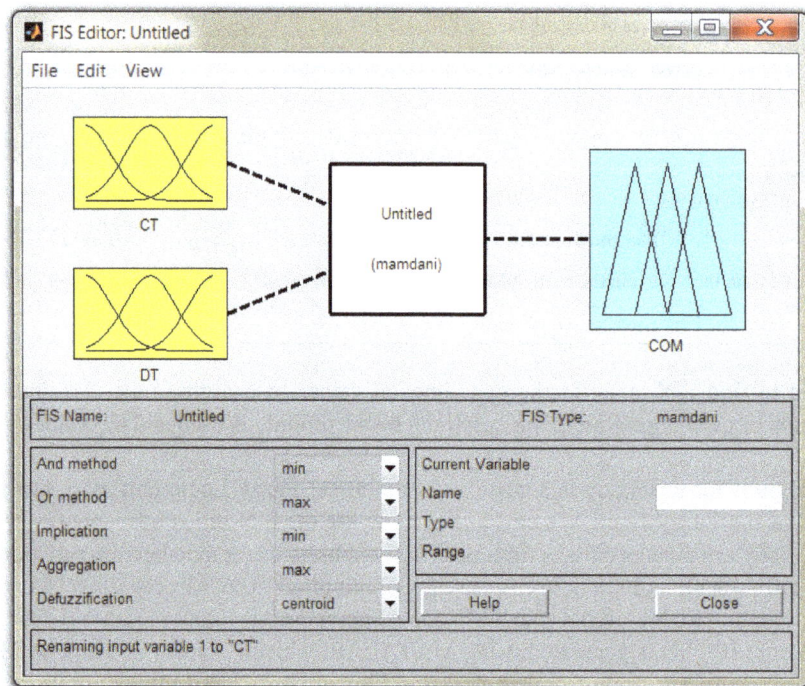

(a)

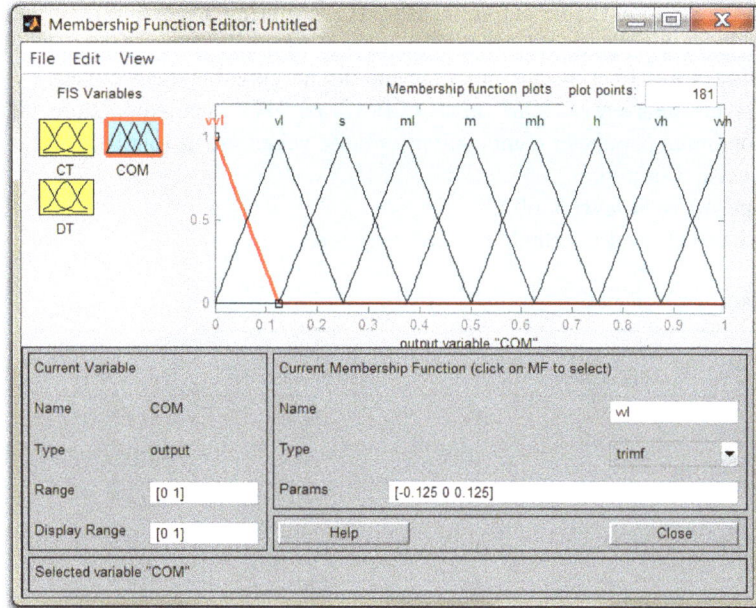

(b)

Figure 10. (a) Two input-one-output Fuzzy Logic unit and membership functions; (b) Membership functions for COM.

Table 6. COM for parameters selection.

Sl. no.	Flow rate	HTF inlet temperature	Capsule shape	Comprehensive output measure (COM)
1	2	60	1	0.9615
2	2	60	2	0.3462
3	2	60	3	0.5283
4	2	58	1	0.6455
5	2	58	2	0.1221
6	2	58	3	0.4266
7	2	56	1	0.5993
8	2	56	2	0.1221
9	2	56	3	0.4266
10	4	60	1	0.9572
11	4	60	2	0.4100
12	4	60	3	0.6933
13	4	58	1	0.6933
14	4	58	2	0.2304
15	4	58	3	0.4266
16	4	56	1	0.6455
17	4	56	2	0.2304
18	4	56	3	0.4266

(ML), medium (M), medium high (MH), high (H), very high (VH) and very very high (VVH) grade. Then, 9 fuzzy rules are defined and shown in **Table 7**.

After analyzing the comprehensive output measure (COM) using Taguchi a rank will be obtained. The rank which is shown in **Table 8** indicates the effect of input parameters on the responses *i.e.*, the rank-1 indicates that the capsule shape has more effect on charging time and discharging time respectively.

Figure 11 represents the average COM values of flow rate, HTF inlet temperature and capsule shape and it is observed that the optimum parameter combination for these values are as follows.

Flow rate at level 2: 4 lit/min.

HTF inlet temperature at level 3: 60°C.

PCM capsule shape at level 1: cylindrical capsule shape.

Table 7. Rules for process parameter.

Sl. no.		S/N ratio for charging time		S/N ratio for discharging time		Comprehensive output measure (COM)
1	IF	L	AND	L	THEN	VVL
2	IF	L	AND	M	THEN	VL
3	IF	L	AND	H	THEN	L
4	IF	M	AND	L	THEN	ML
5	IF	M	AND	M	THEN	M
6	IF	M	AND	H	THEN	MH
7	IF	H	AND	L	THEN	H
8	IF	H	AND	M	THEN	VH
9	IF	H	AND	H	THEN	VVH

Table 8. Selection of parameters.

Level	Flow rate	HTF inlet temperature	Capsule shape
1	0.4642	0.4084	0.7504
2	0.5237	0.4241	0.2435
3	---------	0.6494	0.4880
Delta	0.0595	0.2410	0.5069
Rank	3	2	1

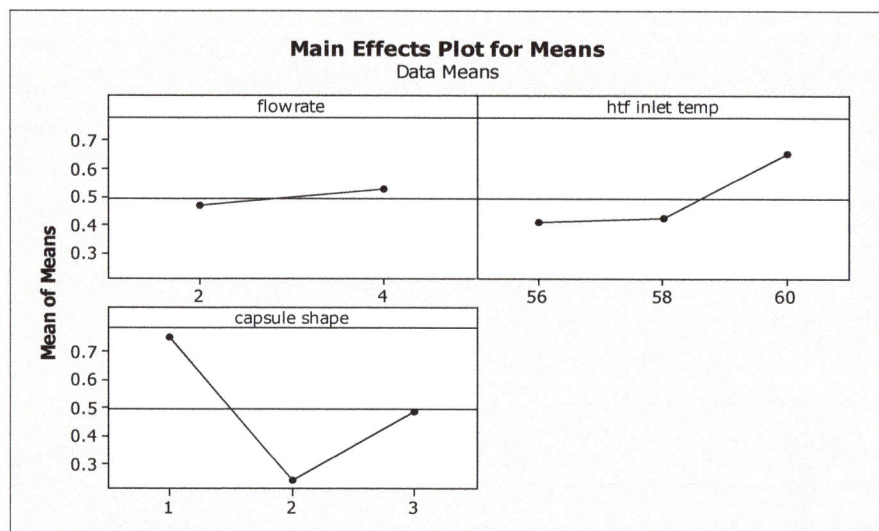

Figure 11. COM value for each factor at each level.

Table 9. Confirmation test results.

	Flow rate (lit/min)	T_{HTF} (°C)	Shape of the capsule	Charging time (min)	Discharging time (min)
Fuzzy logics	4	60	1	6	138
Experimental values	4	60	1	8	140

5. Conclusions

From the results the following conclusions have been drawn:
- The effect of mass flow rate of heat transfer fluid at 4 lit/min and heat transfer fluid inlet temperature at 60°C is more on charging time when compared to other. Hence, it is concluded that higher flow rates and higher inlet temperaStures of heat transfer fluid are recommended.
- From the results, it is also observed that the total energy stored and energy retrieval time are high in combined Sensible heat storage (SHS) and Latent heat storage (LHS) system than conventional sensible heat storage system (SHS). Hence, combined SHS and LHS are recommended for thermal energy storage systems.
- The charging time, surface area to volume ratio and energy retrieval time is more for cylindrical PCM capsule shape compared to others. Hence, cylindrical PCM capsule is recommended for filling PCM.
- From **Table 9**, it is observed that the experimental results and fuzzy optimized results are quite close to each other. Hence, it is concluded that Fuzzy Logics technique can be efficiently applied to optimize the phase change materials parameters.

Acknowledgements

Authors would like to acknowledge the technical staff of Mechanical Engineering Department, S. V. University, Tirupati for their help in this experimentation.

References

[1]   Regin, F., Solanki, S.C. and Saini, J.S. (2009) An Analysis of a Packed Bed Latent Heat Thermal Energy Storage System Using PCM Capsules: Numerical Investigation. *Renewable Energy*, **34**, 1765-1773. www.elsevier.com/locate/renene http://dx.doi.org/10.1016/j.renene.2008.12.012

[2]   Singh, A., Datta, S., Mahapatra, S.S., Singha, T. and Majumdar, G. (2011) Optimization of Bead Geometry of Submerged Arc Weld Using Fuzzy Based Desirability Function Approach. *Journal of Intelligent Manufacturing*, **24**, 35-44. http://dx.doi.org/10.1007/s10845-011-0535-3

[3]   Sharma, A. and Chen, C.R. (2009) Solar Water Heating System with Phase Change Materials. *International Review of Chemical Engineering (I.RE.CH.E.)*, **1**, 297-307.

[4]   Sharma, A., Tyagi, V.V., Chen, C.R. and Buddhi, D. (2007) Review on Thermal Energy Storage with Phase Change Materials and Applications. *Renewable and Sustainable Energy Reviews*, **13**, 318-345

[5]   Kanimozhi, B, Ramesh Bapu, B.R. and Sivashanmugam, M. (2010) Enhancement of Solar Thermal Storage System Using PCM. *National Journal on Advances in Building Sciences and Mechanics*, **1**, 48-52.

[6]   Maheswari, U. and Reddy, R.M. (2013) Thermal Analysis of Thermal Energy Storage System with Phase Change Material. *International Journal of Engineering Research and Applications (IJERA)*, **3**, 617-622.

[7]   Vikram, D., Kaushik, S., Prashanth, V. and Nallusamy, N. (2006) An Improvement in the Solar Water Heating Systems Using Phase Change Materials. *Proceedings of the International Conference on Renewable Energy for Developing Countries*, Denver, 8-13 July 2006, 409-416.

[8]   Oroa, E., Gila, A., Miroa, L., Peiroa, G., Alvarezb, S. and Cabezaa, L.F. (2012) Thermal Energy Storage Implementation Using Phase Change Materials for Solar Cooling and Refrigeration Applications. *Energy Procedia*, **30**, 947-956.

[9]   Rajyalakshmi, G. and Venkata Ramaiah, P (2013) Optimization of Process Parameters of Wire Electrical Discharge Machining Using Fuzzy Logic Integrated with Taguchi Method. *International Journal of Scientific Engineering and Technology*, **2**, 600-606.

[10]   Lu, H.S., Chen, J.Y. and Chung, Ch.-T. (2008) The Optimal Cutting Parameter Design of Rough Cutting Process in Side Milling. *Journal of Achievements in Materials and Manufacturing Engineering*, **29**, 183-186.

[11] Tanishita, I. (1970) Present Situation of Commercial Solar Water Heater in Japan. *Proceedings of the ISES Conference*, Melbourne, 2-6 March 1970, Paper No. 2/73, 67-78.

[12] Archibald, J. (1999) Building Integrated Solar Thermal Roofing Systems History, Current Status, and Future Promise. *Proceedings of the Solar Conference*, American Solar Energy Society, American Institute of Architects, 95-100.

[13] Chandrasekaran, M. and Devarasiddappa, D. (2012) Development of Predictive Model for Surface Roughness in End Milling of Al-SiCp Metal Matrix Composites Using Fuzzy Logic. *International Scholarly and Scientific Research & Innovation*, **6**, 928-933.

[14] Fatih Demirbas, M. (2006) Thermal Energy Storage and Phase Change Materials: An Overview. *Energy Sources, Part B*, **1**, 85-95.

[15] Mehling, H. and Cabeza, L.F. (2008) Heat and Cold Storage with PCM. An up to Date Introduction into Basics and Applications. Springer, Berlin.
http://www.springer.com/chemistry/industrial+chemistry+and+chemical+engineering/book/978-3-540-68556-2

[16] Ercan Ataer, O. (2006) "Storage of Thermal Energy, in Energy Storage Systems" in Encyclopedia of Life Support Systems (EOLSS), Developed under the Auspices of the UNESCO. Eolss Publishers, Oxford. http://www.eolss.net

[17] Karthikeyan, R., Jaiganesh, S. and Pai, B.C. (2002) Optimization of Drilling Characteristics for Al/SiCp Composites Using Fuzzy/GA. *Metals and Materials International*, **8**, 163-168.
http://dx.doi.org/10.1007/BF03027013

[18] Khot, S.A., Sane, N.K. and Gawali, B.S. (2011) Experimental Investigation of Phase Change Phenomena of Paraffin Wax inside a Capsule. *International Journal of Engineering Trends and Technology*, **2**, 67-71.

[19] Vijay Padmaraju, S.A., Viginesh, M. and Nallusamy, N. (2008) Comparative Study of Sensible and Latent Heat Storage Systems Integrated with Solar Water Heating Unit. *International Conference on Renewable Energy and Power Quality*, Santander.

[20] Hasenöhrl, T. (2009) An Introduction to Phase Change Materials as Heat Storage Mediums. Project Report, 160 Heat and Mass Transport, Lund.

[21] Bhatt, V.D., Gohil, K. and Mishra, A. (2010) Thermal Energy Storage Capacity of Some Phase Changing Materials and Ionic Liquids. *International Journal of ChemTech Research CODEN (USA): IJCRGG*, **2**, 1771-1779.

3

A Motor Management Strategy for Optimising Energy Use and Reducing Life Cycle Costs

V. Dlamini, R. C. Bansal, R. Naidoo

Department of Electrical Electronic & Computer Engineering, University of Pretoria, Pretoria, South Africa
Email: muzid@tuks.co.za

Abstract

With increasing energy costs and renewed focus on using energy in ways that support the environment, a structured approach is required to ensure that energy is used efficiently. A comprehensive motor management strategy to reduce motor life cycle costs while increasing reliability is presented. The application of energy management principles is combined with benefits that can be obtained from using energy-efficiency motors. An economic model for determining the optimal time a motor should be replaced with a higher efficiency motor is proposed. The strategy presented incorporates benefits that can be obtained from using in-situ motor efficiency estimation and condition monitoring techniques as part of a motor management system.

Keywords

Motor Efficiency; Motor Management; Maintenance; Vibration Signature Analysis; Energy Efficiency; Motor Replacement

1. Introduction

Electric motors are a key part of industry. They are used in a wide variety of equipment and processes. This includes fans, pumps, compressors, conveyor drives and machine tools. Motors are a leading power consumer due to their widespread use in industry. Motors can account for more than two thirds of the electrical power consumption in some countries. As the cost of electricity continues to increase, motors provide a great opportunity to reduce energy consumption.

Energy conservation technologies can reduce the energy consumption by an estimated 11% to 18% [1]. The reduced energy consumption results in a reduction in operating costs for businesses. This means less power has to be generated which, in turn reduces the harmful greenhouse gases emitted into the atmosphere.

Motor management can be described as strategies that focus on reducing the total cost of ownership of motors in a plant. The cost of ownership of motors includes the energy cost of running a motor, the cost of purchasing motors, the cost of maintaining motors, and the business cost incurred as a result of motor-related process interruptions [2]. A comprehensive motor management strategy incorporates the benefits of the latest technology and the application of best practices to the repair of motors in a plant. The ultimate aim of a motor-management strategy is to ensure a reliable plant at the lowest possible motor-related costs [3].

In this paper an overview of energy management is presented with a focus on electric motors. The benefits that can be obtained from using energy efficient motors are discussed. The cost savings that can be obtained through developing a structured approach to motor management are investigated. An economic model for determining the optimal time a motor should be replaced with a high efficiency motor is presented. The application of non-intrusive motor efficiency estimation and condition monitoring techniques is proposed as part of an integrated real time motor management system.

2. Energy Management Overview

An energy management program must seek to minimize the adverse impact on the environment. This can be achieved by understanding how the business uses energy and creating an awareness of energy saving. Efficient maintenance structures must be put in place [4].

Motor energy management strategies focus on load management, efficiency management and power factor correction [5]. A starting point for energy management is to perform an energy audit. This determines how power is consumed by the plant. Once an audit has been conducted, energy-saving opportunities can be identified. Plans for implementing them can be put in place. The energy audit identifies the following areas for potential improvements [4].
- The efficiency of the operations.
- The efficiency of the billing systems.
- Efficiency of the maintenance activity.

The efficiency of operations entails assessing the design and operation of the different processes in the plant to determine if they use energy efficiently. A motor energy audit must focus on the motor sizes and determine how well they are matched to the load requirements. Incorrect motor sizing has a negative impact on the efficiency of the motor. A motor that is larger than required results in operation at a lower efficiency. This translates to energy loss. Motors usually operate at their highest efficiency at between 75% to 80% of their rated load. The efficiency and power factor both decline as the load reduces. Motors are often oversized to allow for higher future loads or to make provision for short-term load requirements [3].

The billing system must be analyzed to ensure that the economic tariff structure is optimal while considering the plant's operational requirements. It is also important to determine the contribution made by motors to the overall energy consumption of a plant. This allows for the calculation of potential savings which can be realized through motor management strategies. The maintenance activity within the plant has to be assessed to determine the standard. Poor maintenance results in a reduction in efficiency.

A portable instrument for measuring and logging of motor load profiles and estimating efficiency is a valuable tool to conduct an energy audit. It could further be used to determine the power factor of the motor.

Once the data from an audit has been collected, it must be analyzed and opportunities for energy savings should be identified. Action plans must then be put in place. The following alternatives can be implemented:
a) The motor can be kept intact.
b) The motor can be replaced with a new standard motor.
c) The motor can be replaced immediately with a higher efficiency motor.

Control systems such as flux optimization or variable speed drives can be implemented to improve the efficiency of the motor [5].

The action taken depends on the load or process requirements. After analyzing the plant processes, opportunities for optimizing the process efficiency through implementing a control strategy can be identified. A thorough economic comparison of the available options is necessary to maximize savings.

3. Motor Replacement

In this section the development of a motor replacement strategy is discussed. Tools for economic analysis of the potential benefits of replacing motors with energy efficient motors are presented.

3.1. Motor Replacement Strategy

The cost and environmental benefits of replacing a standard efficiency motor with an energy-efficient motor have been highlighted. A motor replacement strategy has to be developed to achieve the benefits. This ensures that clear guidelines exist on how to ensure that motors are operated with the desired reliability and at optimal life cycle costs. An installed motor can be replaced with an energy efficient motor under the following conditions [6]:

a) When a motor has failed;
b) When a new motor is required for an application; and
c) When a motor currently in operation is to be replaced.

In each of the cases an economic evaluation of the available options has to be performed to quantify the benefits. A repair/replacement strategy has to consider the following:

- The impact on energy usage;
- The cost of the capital to be spent;
- The motor size;
- The motor repair cost;
- The motor operating and repair history;
- The replacement motor cost; and
- The availability of a replacement [7].

The energy usage and efficiency of the installed motor must be compared to an energy-efficient replacement motor. An energy-efficient motor provides an opportunity to reduce the cost of energy for operating the plant. There is an opportunity to assess if the motor is properly sized for the load. An oversized motor operates at a lower efficiency level. This results in energy wastage. The repair cost can be used to make the repair/replace decision. If the motor repair cost exceeds a certain percentage of the replacement cost, the motor must be replaced with an energy-efficient motor. The motor operating and repair history is an important factor when making the decision. A strategy for repairing/replacing a motor must consider the reliability of the motor and the probability of future failures. If a motor has been repaired for a predetermined number of times, it should be scrapped. The availability of an energy-efficient replacement motor needs to be investigated. If there is a long lead time for the replacement motor, then the production losses incurred until replacement might be excessive. The lead time for the motor repair has adverse consequences if there are no spares and it runs critical process equipment. The above-mentioned factors need to be taken into account when doing an economic evaluation.

New installations present a good opportunity to introduce energy-efficient motors on a plant. The plant will yield benefits of using energy-efficient motors. The motor strategy for a plant must specify that all new motor installations use energy-efficient motors. A detailed analysis can be done to determine the feasibility of introducing an energy efficient motor as a replacement for a standard-efficiency motor that is still operational. A proposal for how motor replacement decision should be made is presented after a discussion on how motor maintenance influences such decisions.

3.2. Economic Analysis

In order to replace a standard efficiency motor with an energy-efficient motor, a capital investment is required. Before a capital investment is made, an economic analysis has to be performed to determine the return on investment. The return on investment is used to determine the economical feasibility of purchasing a new motor. A challenge in implementing a motor replacement or repair strategy is that the financial benefits of the investment may only be realised a few years later. The justification for the capital investment has to be made at the time the motor is replaced. Methods for performing the required economic analysis are presented in this section.

When comparing different economic investment options, it is necessary to convert them to a common base. Numerous techniques can be used to enable such a comparison. The most widely used methods for enabling economic comparison are the payback, net present value, internal rate of return, project balance and annual equivalent methods. Although all of the tools mentioned can be used, the preferred methods are the net present value and payback methods. These two methods and their application to motor comparison are investigated in detail [4].

To evaluate as to whether replacing a motor with an energy-efficient motor is feasible, information on the process and motors is required. The electricity tariff structure, annual motor load profiles and motor efficiency curves are required to determine the annual power consumption of the motors to be compared. In a plant where there is an established energy-management structure, motor load profiles for each motor may exist. Where such information is not readily available, a power-logging instrument can be used to determine the motor load profile.

The efficiency estimation technique presented in [8] can be used to determine the motor efficiency curve at each point of the motor load cycle. The efficiency curve of the energy-efficiency replacement motor can be requested from the motor manufacturer. The effective interest, energy cost inflation rate, cost of motor replacement and its expected operating life are additional information required to determine the net present value of the investment.

The annual savings that will be realised by replacing a motor with an energy-efficient motor are given by (1), [9].

$$A_{saving} = P_{out} L h_r C \left(\frac{1}{E_c} - \frac{1}{E_r} \right)$$
(1)

where P_{out} is the motor rated power in kW, L is the percentage of full load divided by 100, h_r is the annual motor operating hours, C is the average energy cost per kWh, E_c is the efficiency of the motor currently installed and E_r is the efficiency of the energy-efficient replacement motor. In (1) it is assumed that the motor will operate at the same load when it is in service. In order to be able to compare the total savings that will be achieved, the net present value of the savings have to be calculated over the motor's expected operating life.

Once the annual savings have been determined, the net present value of the savings that will accrue over the motor's operating life can be determined. The net present value is a method that is used to bring the savings that will be realised over an extended period to their present equivalent. It takes the time value of money [4] into account. This calculation must include the cost of the capital required to purchase the new motor and a projection of the expected inflation rate for the cost of energy over the operating life of the motor. To determine the present value of the savings, it is necessary to calculate the effective interest rate using (2).

$$i = \frac{100 + r_2}{100 + r_1} - 1$$
(2)

where i is the effective interest rate, r_1 is the expected annual energy cost inflation rate and r_2 is the required internal rate of return on investments. The inflation rate and internal rate of return are assumed to be constant over the calculation period. After the effective interest rate has been calculated, the present value of the savings to be obtained can be determined from (3).

$$NPV_{saving} = A_{saving} \frac{(1+i)^n - 1}{i(1+i)^n}$$
(3)

where A_{saving} is the annual savings, i is the effective interest rate and n is the expected operating life of the new motor. The present value of the savings obtained can then be compared to the expense that will be incurred in purchasing the new motor. If the cost of the new motor is less than the net present value of the savings that will be achieved, a business case can be presented.

In providing economic justification for motor replacement, the payback period can be used as an alternative method for making the decision. The payback period for a motor replacement study is the time it will take for the benefits of replacing the current motor to exceed the capital invested in purchasing the new motor. The payback period can be calculated using (4), [9].

$$n_{PB} = \frac{\ln \left(- \frac{A_{saving}}{iC_{motor} - A_{saving}} \right)}{\ln(1+i)}$$
(4)

where n_{PB} is the payback period in years, A_{saving} is the annual savings, i is the effective interest rate and C_{motor} is the cost of the replacement motor. The cost of the new motor must include labour and downtime for the installation and uninstalling. When using the payback period for deciding the feasibility of replacing a motor, a project with a shorter payback period is most feasible.

In determining the annual savings it should be noted that Equation (1) is only applicable to motors that operate under constant load. This is because the efficiency of both motors under consideration will vary with different loading points. The annual savings do not take into account the demand charge for electricity. In cases where the motor is considered to make an appreciable contribution to the maximum demand, the annual energy savings calculation has to be modified to take this into account [10]. The power saved in kW can be calculated using (5).

$$P_{saving} = P_{out}L\left(\frac{1}{E_c} - \frac{1}{E_r}\right) \qquad (5)$$

The annual savings that will be achieved by reducing the maximum demand are determined using (6).

$$D_{savings} = 12 \times P_{saving} \times D_c \qquad (6)$$

where P_{saving} is the power saving in kW and D_c is the demand charge. The total annual saving, T_{saving} is given by (7). The total annual saving can be substituted for the annual savings in the expression for calculating the net present value and the payback period:

$$T_{saving} = A_{saving} + D_{savings} \qquad (7)$$

If the motor load profile is not constant the equations presented need to be applied to each relatively constant portion of the motor load cycle that is relatively constant.

An example of such a load profile is shown in **Figure 1**.

The annual savings are determined as follows:

1. The calculation would be done for each of the three loading points, A, B and C. The annual operating hours would be determined by multiplying the daily hours at each loading point by the number of days the motor operates per year.
2. The output power for each of the loading points would be determined.
3. The annual saving would be calculated using the efficiency at each loading point.
4. The total annual saving would be determined using (8) as the sum of the savings for each loading point:

$$A_{saving} = \sum_{A}^{C} P_{out}Lh_rC\left(\frac{1}{E_c} - \frac{1}{E_r}\right) \qquad (8)$$

In developing a motor replacement policy the tools that have been presented can be applied. It is recommended that minimum economic requirements be determined and adopted for a company. This guides decision making on motor replacement based on either the net present value or the payback period. A decision should not only be based on motor replacement cost. It must factor in the expected future repair costs of the older motor installed in the plant. A comprehensive economic analysis must also explore the benefits obtained from power factor correction and the impact of available rebate programmes for energy-efficient motors. This will form an integral part of a motor-management strategy.

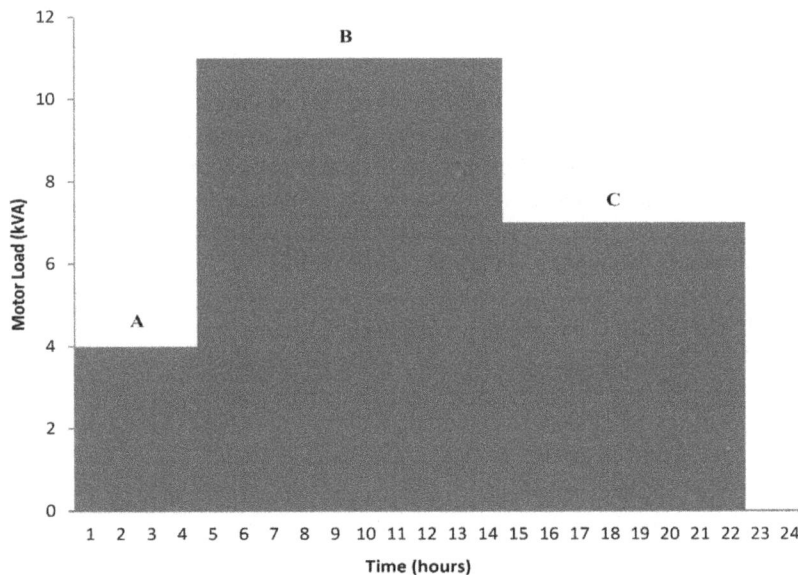

Figure 1. Motor load daily load profile.

4. Motor Maintenance

The importance of a good maintenance strategy in an organization is often understated. **Table 1** shows the contribution made by maintenance-related to the total operating costs [11]. This varies across different industry sectors. The maintenance cost is made up of the following components.

a) The direct cost of repairs
b) The cost of any pro-active work (labor, materials, contractor, etc)
c) The cost of lost revenue and reputation due to downtime
d) The cost of any penalties that are incurred as a result of damaged products or operating systems

It is can be seen from **Table 1** that the cost of maintenance can be too significant to ignore in any plant. A good motor maintenance program extends the life of motors and improves system availability. This will result in a reduction in maintenance costs and downtime-related losses [3]. A motor-maintenance strategy must have guidelines for motor storage, installation, operation and repair.

Motors that are kept in storage for have a higher probability of failure if they are not stored according to best practices. The way in which a motor is installed has a significant impact on its reliability and operating life. When a motor is installed it is essential to ensure that is has the proper foundation and alignment procedures in place to minimize additional stresses to the motor. This can exceed the design limits [3]. Motors should be operated according to the manufacturers' guidelines to ensure that they achieve the design life.

A reliability-centred maintenance (RCM) approach has been found to be effective for motors [12]. RCM is a proactive maintenance strategy that has processes for anticipating which failure modes will occur [11]. Once the failure modes have been determined the consequences of each failure mode are analyzed. Plans are developed to eliminate or minimize the consequences of each of the failure modes.

The repair of motors presents an opportunity to implement policies that will result in long-term savings. The repair of motors must be governed by guidelines to ensure repairs are only carried out when it is financially feasible. Managing motor repairs starts with establishing repair specifications and identifying suitable suppliers that can provide high-quality repairs [7]. Agreements must be put in place with supplies to ensure that the repair of motors is done according to industry best practices. The motor repair decision flowchart in **Figure 2** illustrates how decisions can be made when a motor fails in order to minimize the life cycle costs. Keeping a detailed motor repair history is important. Analyzing the trends can allow for estimating the life of the motor based on its reliability and age. This can prevent the repair of a motor that is near the end of its life and allows for the introduction of an energy-efficient motor in its place. The proposed decision-making process focuses on using all available data to make decisions that will realize cost savings over the operating life of the motor and plant. It guides a motor manager to take a holistic approach that takes into account energy efficiency and long-term benefits.

5. Motor Management Strategy

The discussions of the preceding sections lead to the development of the proposed motor-management strategy to ensure that motors in a plant have the lowest possible operating cost and very high reliability. A total motor-management approach with real time efficiency, load factor, power factor and vibration-based condition monitoring is proposed. The collected date will be stored on a database for use in the RCM based approach. With the availability of the relevant data, the RCM process should provide a comprehensive preventative maintenance program that will ensure that motors operate at high reliability.

Table 1. Contribution of maintenance to operating costs.

Industry	Contribution (%)
Mining	20 - 50
Primary metals	15 - 25
Electric utilities	15 - 25
Manufacturing	5 - 15
Processing	3 - 15
Fabrication and assembly	3 - 5

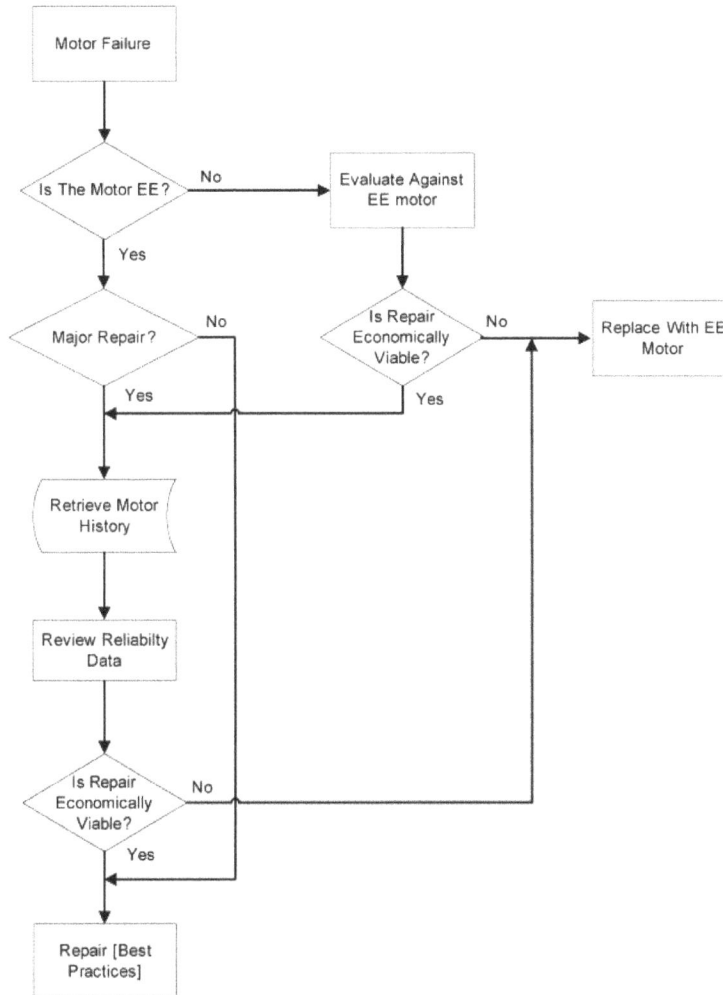

Figure 2. Motor repair decision flowchart.

An accurate method of estimating the efficiency of in-service motors is needed in order to determine the performance of installed motors without disrupting the motor driven process. In the proposed system the motor efficiency is estimated using a non-intrusive implementation of the compensated slip method. The motor speed is accurately estimated using motor vibration signature analysis [8]. The estimated efficiency can then be used to calculate the annual savings that will be realized by replacing an installed motor with an energy-efficient motor using (1). If the load in not constant than potential savings it can be obtained with (8). This can be incorporated into a motor management support system as described below.

The vibration signal may also used for condition monitoring since motor vibration signature analysis has been widely studied and applied in this field. Various types of motor faults can be detected from the vibration signature of a motor. These include faults such as winding faults, unbalanced stator and rotor parameters, broken rotor bars, eccentricity and bearing faults. The faults are detected as harmonics using vibration signature analysis [13]. The condition monitoring data should be used as input data in the RCM system. It provides the data required to successfully implement the replacement decision model in **Figure 2**.

A motor management support system that will make implementing the motor-management strategy easier and more efficient should be used. In **Figure 3** the proposed plant wide motor management system is shown. This system consists of local field instruments at each motor for measuring its voltage, current and vibration. The collected data are sent via a wireless network to a central processor. At the central processor the non-intrusive speed and efficiency estimation technique is implemented [8].

Energy usage and reliability indicators for each motor can also be calculated at the central server.

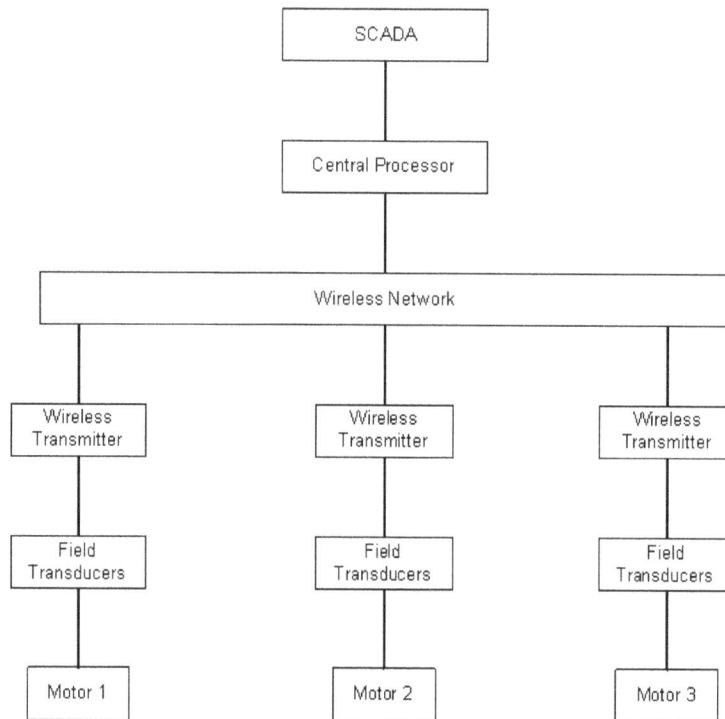

Figure 3. Motor management support system.

The proposed system should include the vibration based condition-monitoring tools with the fault detection algorithms implemented on the centralised processor. The processed data are sent to a Supervisory Control and Data Acquisition (SCADA) system so that historic data can be accessed for use in motor replacement and energy-management studies. Alarms can be sent to the SCADA system to inform control room operators or maintenance personnel of developing motor faults so that corrective action can be planned and scheduled. This will result in cost saving because unplanned downtime will be reduced.

6. Conclusions

An overview of motor energy management has been presented. Tools for performing an economic comparison of motors were discussed. The tools can be used to determine the annual savings that can be realized by using an energy-efficiency motor, as well as the payback period on the initial capital investment. The non-intrusive compensated slip method can be used to estimate the efficiency of an installed motor. This enables the above-mentioned calculations to be made.

The motor speed estimation technique based on vibration signature analysis can be used for detecting numerous fault conditions. A method for deciding how to handle motor failures has been proposed. This method takes into account the economic aspects of repairing the motor, whilst considering the reliability history of the motor. It provides a guideline for how energy efficient motors can be introduced taking into account the life cycle costs including the cost of energy.

A comprehensive motor-management strategy is proposed. This motor management strategy centres on finding the correct balance for energy efficiency, motor reliability and maintenance costs. This is achieved by using a motor-management system built around a network of field sensors to provide real-time data on the condition and performance of motors in a plant. This allows for proactive decisions to be made to minimize breakdowns and optimize energy usage.

References

[1]   Saidur, R. (2010) A Review on Electrical Motors Energy Use and Energy Savings. *Renewable and Sustainable Energy Reviews*, **14**, 877-898. http://dx.doi.org/10.1016/j.rser.2009.10.018

[2]   Mason, I. and Jones, T. (2004) Proactive Motor Management Can Help Reduce Operating Costs in the Pulp & Paper Industry. *Proceedings of the Pulp and Paper Industry Technical Conference*, Boston, 27 June-1 July 2004, 70-72.

[3]   Basso, D., Nyberg, C. and Yung, C. (2007) The Repair/Replace Decision from a Total Motor Management Perspective. *Proceedings of the Pulp and Paper Industry Technical Conference*, Williamsburg, 24-28 June 2007, 235-241.

[4]   Capehart, B.L., Turner, W.C. and Kennedy, W.J. (2012) Guide to Energy Management, Energy Management. 7th Edition, The Fairmont Press, Lilburn.

[5]   Li, Y. and Yu, H. (2007) Energy Management for Induction Motors Based on Non-Intrusive Efficiency Estimation. *Proceedings of the International Conference on Electrical Machines and Systems*, Seoul, October 2007, 1763-1766.

[6]   Akbab, M. (1999) Energy Conservation by Using Energy Efficient Electric Motors. *Applied Energy*, **64**, 49-158.

[7]   Whelan, C. Sassano, E. and Kelley, J. (2004) Management of Electric Motor Repair. *Proceedings of the Petroleum and Chemical Industry Technical Conference*, Wilmington, 13-15 September 2004, 279-288.

[8]   Dlamini, V., Naidoo, R. and Manyage, M. (2013) A Non-Intrusive Method for Estimating Motor Efficiency Using Vibration Signature Analysis. *International Journal of Electrical Power and Energy Systems*, **45**, 384-390. http://dx.doi.org/10.1016/j.ijepes.2012.09.015

[9]   Phumiphak, T., Kedsoi, T. and Chat-Uthai, C. (2005) Energy Management Program for Use of Induction Motors Based on Efficiency Prediction. *Proceedings of IEEE Region* 10 *Conference*, Melbourne, 21-24 November 2005, 1-6.

[10]   Eltom, A. and Aziz, M.A. (2005) The Economics of Energy Efficient Motors during Unbalanced Voltage Conditions. *Proceedings of IEEE PES Conference and Exposition in Africa*, Durban, 11-15 July 2005, 378-384.

[11]   Campbell, J.D. and Reyes-Picknell, J.V. (2006) UPTIME Strategies for Excellence in Maintenance Management. 2nd Edition, Productivity Press, New York.

[12]   Penrose, H.W. (2005) RCM-Based Motor Management. *Proceedings of Electrical Insulation Conference and Electrical Manufacturing Expo*, Chesapeake, 26 October 2005, 187-190.

[13]   Singh, G.K. and Kazzaz, S.A.S.A. (2003) Induction Machine Drive Condition Monitoring and Diagnostic Research—A Survey. *Electric Power Systems Research*, **64**, 145-158. http://dx.doi.org/10.1016/S0378-7796(02)00172-4

4

The Future Role of the Nuclear Energy in Brazil in a Transition Energy Scenario

Fábio Branco Vaz de Oliveira, Kengo Imakuma, Delvonei Alves de Andrade

Nuclear and Energy Research Institute, Cidade Universitária, São Paulo, Brazil
Email: fabio@ipen.br, delvonei@ipen.br

Abstract

This paper discusses and presents figures about the future power consumption in the world and, especially in Brazil, based on the current world and Brazilian's energy scenarios. Emphasis is given to the scenarios of nuclear power and uranium resources demand. A discussion on the future roles of thorium and uranium fuels in the replacement of the traditional resources like oil and gas is also presented, as it is the role of the new nuclear power plants, planned to be built in a short term time horizon. This paper considers two different indexes for future projections, and the results obtained indicated a strong dependence on them. The time horizon for the analysis was fixed on the time estimated for Brazil to reach its maximum in population, and parameters evaluated were taken from the Brazilian's governmental and world data on the population growth, energy consumption and energy consumption per capita. Calculations show that the power consumption projections for Brazil, for the adopted time horizon and working with global indexes, become overestimated, when compared with the results considering the national indexes. According to our approach, power consumption estimates using global indexes becomes approximately 4.5 times higher than the estimates presented by the Brazilian indexes. This was the motivation to the discussion between the Brazilian and world energy demand scenarios, and also the roles of nuclear energy in the future transition from the current conventional to alternative sources.

Keywords

Nuclear Energy, Renewables, Future Energetic Resources, Uranium, Projections

1. Introduction

The shortage of the most common used energy resources, such as oil, coal and water, and their growing consumption, together with the population growth and global warming, are phenomena clearly affecting the planet. The recent formation of tornadoes in the coast of Santa Catarina, which never happened or had been reported in all Brazilian history, the excessive melting of the polar ice caps, for example, can be foreshadows of a large scale global manifestations of uncontrolled and unpredictable climate changes, due to the exaggeration of the consumption of natural resources. Also, the increasing gap between economical and social classes and the pressure it exerts in the production of more expensive products, which also generates an enhancement in the energy consumption, clearly become an important issue on the countries future energetic planning. Besides the social and economical needs, planning must take into consideration the structure of the national energy matrix, and has to be based on what it can provide to support the country's growing perspectives. In this scenario, studies concerning the use and improvement in the technologies related to renewable or clean energies are growing and nuclear energy, among them, plays an important role.

Energy from renewable and nuclear resources, are considered candidates to reduce an incoming world energy crisis. Nuclear energy is one of the most promising candidates, despite the recent accident in Fukushima, since the technology is well developed and has a strong knowledge basis. Some countries are declining to build new nuclear plants. Nonetheless, it is interesting to note, for example, the acceleration of the construction of more secure power plants and the adequacy of the older ones to the new and more stringent security criteria in the world and in particular in United States. It shows a pressure in the production of energy, at least for this country. This can be an image of what might occur in the world in terms of energy consumption, but in a lesser scale, since United States, together which China, are the countries exerting the highest pressure in the consumption of the global energy resources [1] making their demands for energy more urgent. An argument which favors the utilization of nuclear energy is given, for example, in the work of Haratyk and Forsberg [2]. They devised a coupling between nuclear energy and renewable facilities. According to them, renewables cannot supply the increasing demand of energy in the short term, given the current countries technological status.

Thus, it is expected that nuclear energy will replace in a large scale and in a short term, the conventional resources, increasing its participation in international and national scenarios of energy demand. But it cannot be accomplished without a deep study about the future supply of nuclear fuels, mainly uranium and, for a more advanced perspective, thorium.

Both minerals are abundant in Brazil, and for the national projections, uranium is available in sufficient amounts to supply the four planned PWR reactors, to be constructed up to 2050 [3]. Even having the technology knowledge of all the fuel cycle, the energy matrix in Brazil is mainly hydroelectric and nuclear represents 3%. In the world, nuclear energy is about 17% of the total generated electricity, and about 5.4% of the total world energy consumption. However, in a transition energy scenario, even with the recent discoveries of oil reserves in pre-salt formations in the Brazilian coast, having a potential to increase power in our energy matrix, with the construction of the Belo Monte hydroelectric power facility, abundance of solar radiation over the entire year covering all our territory, etc., nuclear power could be a short term solution of a possible energy supply delay, expected for the years to come, up to the development of the renewables technology to reach efficiency levels, suitable for commercial exploration. Besides, the development of the technology of the extraction and purification of the small amounts of uranium from the recent discovered phosphatic deposits [4], which covers almost one third in area of the Brazilian territory, will improve the reserves and also the capabilities of the expansion of its nuclear and, thus, energetic matrix.

The construction of nuclear power facilities is also a matter of public discussions, mainly concerned about the safety requirements and accident risks, which rises again after the recent Fukushima's accident. From the arguments above and the discussions to be presented in the next items, it is easy to see and, according to our opinion, that Brazil has a privileged position in terms of the availability of energy resources, including uranium. This privileged position could be used to overcome public rejection and to support the acceptance, even in the long term, to the Brazilian's initiative in the direction of the construction of new nuclear power facilities in Brazil, since massive investments in nuclear power, like those in United States and China, are not planned [3]. However, for a long term planning, energy consumption estimates must be calculated on reliable basis, emphasizing both the global and national scenarios in the current energy context.

This paper analyzes and estimates the future power consumption in the world and in Brazil, based on the current worlds and Brazilian's energy scenarios, with emphasis to the scenarios of nuclear power and uranium resources demand. A discussion about the future roles of thorium and uranium fuels in the replacement of the traditional resources like oil and gas, the advantages of thorium in terms of energetic efficiency, is also presented. As a consequence of the uses and availability of the nuclear fuels, the role of the new nuclear power plants, planned to be built in a short term time horizon, is also discussed in this same scenario. Emphasis was given to the called 2DS scenario [5], which means a scenario where it is predicted a minimum of 2°C for the increasing in the global mean temperature, up to 2025. Actions and plans are under discussion, each one of the possible new technologies having specific advantages and drawbacks [5].

The comparison presented in this paper pointed out a strong dependence on the two different indexes, used for future projections. Global indexes were extracted from the projections made in the Tomabechi's work [6] for the world consumption, which deals with the relation between energy consumption indexes and world's population growth as the parameter to be evaluated. These results were extended to make projections for the Brazilian case. Brazilian indexes were based on the national projections for the population growth, listed by IBGE [7] and were used for the future projections. To check for the importance of domestic scenarios in the projections, graphics are given to show the evolution and to compare global and domestic parameters such as population growth, power consumption per capita and total power consumption in some countries.

The time horizon for the analysis was fixed on the time estimated for Brazil to reach its maximum population, and parameters evaluated were taken from the Brazilian's governmental and world data on the population growth, energy consumption and energy consumption per capita.

This was the motivation to both discussions carried out here, one associating Brazil within a global transition scenario, and the other which dissociates our country from the global considerations, discussing our future projection results in a domestic perspective. Since nuclear is the prompt alternative instead of the undeveloped renewables technology, it is also discussed its role as the main alternative to the replacement of the current conventional ones.

2. Nuclear Energy in a Transition Energy Scenario

Concerning today's debates regarding energy demand and global warming, a transition energy scenario, according to reference [5], means that world needs to change from the prevalent high-carbon/high-pollutant technologies to the incoming low-carbon/low pollutant ones, from the exploitation of the energy resources through the generation of energy. However, considering the phenomena of the development of those new technologies as a dynamical process, demanding time and consuming resources (energetic and monetary), it is known that it usually does not evolves with time at the same rate as that necessary to overcome the problems for the today's climate changes and energy consumption.

Since earth is a dynamical system, it is continuously changing, through changes in its observable properties with time. However, since the time human beings are exploring its resources, they become one of the major causes of changes on earth's properties, the most sensitive of them and most easily visualized and studied are those related to weather. The problem is, at the today's rates of consumption, these changes are possibly leading humanity to an irreversible path and, the worst of all, in a short period of time. Technology and human intelligence solved several problems of our everyday life, enabling us to go to the moon, to explore planets, the universe and the subatomic world, to defeat diseases, communication problems, etc., but its side effects is presenting itself as a very harm one.

Based on the current status of the renewable technology development and the growth of the mean earth's temperature, scientists devised 3 possible states for the future global weather, based on 3 possible values for the increase in the mean earth's temperature. Based on that, governments can take action plans to deal with these three possible scenarios; changes in the energy policy are necessary to be planned, defined and studied to provide approaches for the solution of the incoming energy problems. The above scenarios go from the minimum to the maximum harm for the living beings in the planet. They are called XDS scenarios, where X = 2, 4 or 6 and stands for the expected increase in the mean earth's temperature, in degrees C, for a fixed time horizon of 2025.

It is pointed out by reference [5] that energy technologies like solar and photovoltaic are promising to reach the 2DS requirements, but the ones with greatest potential to reduce carbon emissions, like the nuclear, are making the slowest progress, since security is the main topic of interest for the public acceptance and for its full utilization. It is expected a growth of 600 GW of nuclear installed capacity to reach the 2DS requirements, but after the Fukushima's accident it is also expected a deployment in 100 GW, becoming this process a complicated one.

For the projections of the future of the energy supply by nuclear energy, the repercussion of the 2011 accident was fundamental, leading United States to review the current security criteria for the construction of the 22 planned new nuclear power plants, which are based on the intrinsically safe AP1000PWR design, mostly from 100 MW to 300 MW power. Despite of its safety aspects, however, some authors still criticize its design, like Piore [8], who emphasizes that the security factors for the pressure vessel's project is undersized. For the breeder reactors, another technology being considered which enables the extension of the fuel's lifetime, Cochran [9] emphasizes the problem of the isolation of sodium and water lines, critical due to the risk of explosions. Those are issues that testify against the use of such technologies.

However, according to the World Nuclear Association data [10], in addition to the today's 441 nuclear reactors in the world, more 338 nuclear reactors are planned to be constructed, four in Brazil [10]. Based on the same report [10], in 2011 83 reactors were under construction, most of them are PWR, according to **Table 1**.

On June 1st, 2010, Brazil restarts the construction of the Angra III nuclear power reactor and, together with the planned four, as stated recently in 2012. Our supply of nuclear power will increase in approximately 5300 MW. More ambitious construction plans are presented by China, Russia, United States and Ukraine which are planning to build, respectively, 115, 40, 30, 28 and 20 more nuclear power reactors, a total of 223, mostly based on the LWR technologies. Thus, despite the accident and the rising of the unfavorable public opinion against the construction of new nuclear power plants, it is observed [5] that most of the countries kept their nuclear programs active. But the only factor that has changed is the speed on which the new facilities are planned to be constructed.

Another favorable point for the installation of new nuclear plants was devised by WEC [11], also having as basis that this technology is well developed. In this document it was suggested that, to accelerate the transition from Green House Gas, GHG, to non-GHG technologies, nuclear expertise could be shared from the nuclear developed countries to the non-developed ones. According to WEC [11], this could be another way of global scale cooperation for the reduction of GHG emissions and to develop the 3rd world technological infrastructure, aiming the minimization of the time for an undeveloped country to reach an acceptable level of technology to deploy clean energy technologies. Also, [11] states that new rules for patent interchange and utilization must be redefined, for the humankind to adapt to the new survival requirements. Sharing costs could be viewed by the leading technological countries in the world as a logical path. Since global warming and energy consumption is a world problem, large-scale technology as well as costs should be diffused at same basis in order to share a solution in the direction of the reduction of the global costs of the pollution through the deployment, for example, of the clean energy technologies. This is an important point of debate in the more recent seminars on greenhouse gas emissions and the future of the climate of the planet

We have to keep in mind that important differences arise if the scenario for the estimates is taken locally. For example, countries with high number of nuclear reactors, large populations and with large energy consumption, like United States, China, European Union, etc., global indexes obviously do not reflect the truth about the national estimates for these countries, since they pull the mean world values of consumption/demand of electricity upwards. For example, in terms of the greenhouse gas emissions, it is observed that 50% of the total is due to oil and coal usage, mainly from United States and China. Thus, there is a mistake of sharing the indexes/costs. It is suggested here a criterion for the evaluation of a real "contribution" of a country for the global energy consumption.

For Brazil, the technology for clean energy deployment is still in a growing phase for some of the renewables, in terms of the domestic energy supply index. According to Cerri [12], 41% of Brazil's total domestic energy supply comes from renewable resources, against 14% as a mean value in the world and 6% in industrialized countries.

Table 1. Nuclear reactors under construction or "almost so", by design [10].

Reactor Design	Quantity
Fast Breeder Reactor	02
Pressurized Water Reactor	61 (China = 26)
Pressurized Heavy Water Reactor	11
Advanced Boiling Water Reactor	08
High Temperature Gas Reactor	01
Total	83

This outcome implies a reduced dependence for imports of energy sources, and an advantage in terms of energy supply for the future, even its nuclear park representing only 3% of the total energetic production. Thus, the use of local indexes could be a better choice, for a better accuracy for projections of energy consumption. But it is important to observe that Brazil holds the position to be one of the top green-house gas emitters in the world, mainly due to the deforestation to livestock and agricultural uses [12].

3. The Uranium Scenario

To see how uranium resources can provide a short term solution to face a possible energy supply crisis, it is convenient to compare the world's uranium reserves capabilities with those of oil and coal, the most used natural resources in the world.

For generating 375 GWe, the corresponding electricity generated by nuclear fissions, it is required 68,000 t of uranium per year. According to **Table 2** [13], and ignoring the so called secondary sources of uranium (uranium from nuclear weapons, reprocessing, etc.), if the total estimated amount of uranium metal is 5.5 Mt, at the above rate of consumption the uranium reserves will last 80.8 years. This amount is enough to reach the so called 2DS objectives up to 2025 [5], where the required power related to nuclear energy is about 600 GWe. The difference of 225 GW corresponds to 40,000 t of uranium per year, an enough amount to cover 2DS requirements. As explained below, the Fukushima accident changed some policies related to the construction and operation of new nuclear power plants, the value of 600 GWe must be reduced for this reason.

In terms of oil, based on the current consumption rate, 8.2×10^7 barrels/day, and considering it as constant, a non-realistic hypothesis, its length can be calculated in almost 41 years. United States is the planet's largest consumer of this resource, with approximately 1.9×10^7 barrels/day, nearly 25% of the world's consumption, and Brazil appears in seventh place, 2.5×10^6 barrels/day. Regarding coal, China is in the first place, with 1.31×10^9 tons/year, followed by the United States, with 1.06×10^9 tons/year, and together account for 50% of the world's total consumption, and also for the corresponding greenhouse gas emissions. Brazil appears in 19th

Table 2. Known recoverable resources of uranium (Reasonably Assured Resources plus Inferred Resources, to US$ 130/kg U, 1/1/09, from OECD NEA & IAEA [10]).

Country	Tons of uranium	% of world's total
Australia	1,673,000	31.0
Kazakhstan	651,000	12.0
Canada	485,000	9.0
Russian Federation	480,000	9.0
South Africa	295,000	5.5
Namibia	284,000	5.0
Brazil	279,000	5.0
Niger	272,000	5.0
United States	207,000	4.0
China	171,000	3.0
Jordan	112,000	2.0
Uzbekistan	111,000	2.0
Ukraine	105,000	2.0
India	80,000	1.5
Mongolia	49,000	1.0
Other countries	150,000	3.0
World total	5,404,000	100%

place, with 2.3×10^7 tons/year, twice orders of magnitude less than the previous two [14]. In terms of proved coal reserves, considering that there are around 522 billion tons, the current consumption rate of 4.59×10^9 tons/day [14], and supposing that rate remains constant for the following years, we can estimate that the reserves will last 113 years. In both cases, uranium can be seen as the readiest resource for their possible replacement. For example, as previously mentioned, primary resources of uranium will end, at the current consumption rate, in about 80 years, in the mean value, accounted for the world, not for a particular country figure.

For Brazil, with the estimated growth scenarios for our nuclear matrix [15], from the current 1.95 GW to 3.5 up to 7 GW, our reserves will last 120 to 240 years (reserves of uranium considered at a total cost of less than US\$ 130/kg, the criteria for an economical exploration of an uranium mine) and from 200 to 400 years, taking into account also the inferred reserves. A safer estimate, which predicts a maximum percentage of the total electricity generated by the Brazilian nuclear matrix of 5.7%, excluding the estimated by INB [4] inferred reserves of 800,000 tons, the duration of our uranium resources is, in the worst scenario, about 90 years. Taking into account these reserves, together with the 279,000 ton of uranium metal, the estimates are for the operation of 10 nuclear reactors like Angra III, for 200 years [16]. The problem is that most of the uranium in the inferred reserves comes from phosphates, and the technique for uranium extraction from phosphates is still uneconomical. The same problem in terms of the costs/economy of extraction is presented with the uranium obtained from sea water, whose amount is estimated in 3.3 ppb [6].

Since most of the energy generated in Brazil is from hydropower, and since the water resources are now limited due to considerations of preservation and other environmental aspects [16], to cope with the hydropower restraint in terms of space, nuclear expansion in terms of the self-sustained scenario would be helpful, according to the data presented in **Table 3** presented in the next item.

Thorium is another source of fissionable isotope, U^{233}, its reserves are estimated to be 3 times higher than uranium's. In the work of Ashley [17], it was stated that the amount of thorium as a by-product of rare earth processing would be enough to feed 200 nuclear ADTR ("Advanced Thorium Reactor"), without the need to open new mines for exploration and extraction, thus with no initial investments. Also, other advantage of thorium compared to uranium is its total usage as nuclear fuel as extracted from the mines, instead of the 0.7% relating to the fissile isotope U^{235}, present in U^{238} matrix. It means, according to Ashley [17] that for each GW of electricity produced, only 1 ton of mined thorium is necessary, whereas 200 tons of uranium from the mines or 3.5 Mt of coal would be required to produce the same amount of electricity.

Although there are still doubts about the real need for the construction of new nuclear power plants in Brazil, the debate is still open, and the numbers shown in the tables and the scenarios presented above are favorable. Considering the capabilities of the Brazilian nuclear reserves, it will be described and simulated, in the next items, some possible scenarios for its utilization in relation to two parameters, among the huge numbers of variables that usually affect a more complete analysis, population growth and energy consumption.

The scenarios for the national analysis were taken from the Plano Nuclear Brasileiro 2030 ("Brazilian Nuclear Plan 2030") [18], from EPE and Eletronuclear. A comparison will be made with results found in recent literature.

4. Projections: Energy Consumption and Demand, and Population Growth

Since population and the energy consumption growths are some of the most significant parameters to pressure energy demand, the projections for the Brazilian population growth, presented by the data of IBGE [7] in **Figure 1**, are firstly analyzed before the estimates of energy needs. Other sources for population growth projections are available, like the CIA fact book [1]. According to IBGE, the maximum in Brazilian population will occur in

Table 3. Electricity consumption generated via nuclear power, for inhabitant year, according to the scenarios of the PNB 2030 and using Equation (2).

Scenarios of growth	Year 2030 [MJ/(inhab. year)]	Year 2040 [MJ/(inhab. year)]
Minimum (2.7%)	190.89	200.34
Development (4.2%)	296.94	311.64
Self-sustainable (5.7%)	402.99	422.94

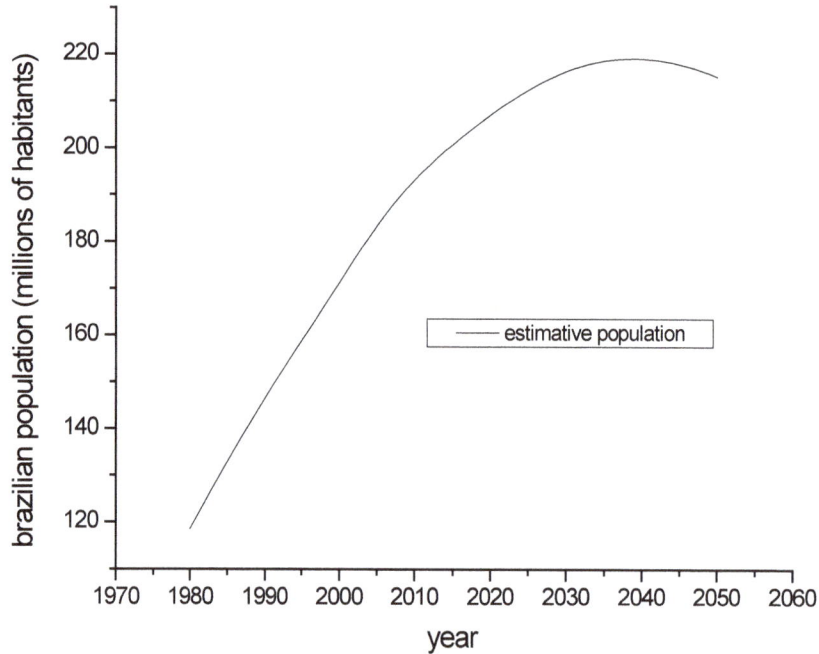

Figure 1. Brazilian population projection, up to 2050 [7].

2039 (219,124,700 inhabitants). However, between 2020 and 2030, the limits of the proposed energetic planning horizons [3] [19], our population will have approximately 212 million inhabitants.

The parameter energy consumption growth, would give accurate predictions if analyzed for each country, instead of taking the estimates based on global indexes. To put both parameters together, we observe that Iceland has the highest index of energy consumption per capita in the world, but its population is below 5 million people, meaning its contribution to the total world energy consumption is low. Countries with the highest energy consumption per capita are China, the United States, Japan, Russia, India, Canada, South Korea, South Africa, Australia, and the European Union. Taking together population growth and absolute population, these countries are not the first ranked ones (in terms of growth rate, East Timor, with 4.5% by year, is in the first place [1]), but pressure the energy consumption by their high number of inhabitants and also by their high energy consumption per capita. The Brazilian energy consumption per capita can be seen in **Figure 2**. The consumption per capita for the most populated countries is given in **Figure 3**, to acquaint for the differences mentioned in the introduction.

In terms of world energy consumption, it is important to observe that countries of the Europe and North America have lower population growth rates, but their energy consumption is about 1 order of magnitude higher than the mean world energy consumption. This is shown in **Figure 4** and **Figure 5**.

In a recent paper from Tomabechi [6], it was estimated that when the world population reaches N_{hm} = 10 billion inhabitants, around the next 30 to 50 years, according to the current global mean population growth of 81 million people/year, the energy demand will be C_m = 2 ZJ/year (where 1 Z = 10^{21}). This number was defined, according to the authors, based on data about the energy resource consumption of developed countries, extended to all the countries of the world. Thus, the estimated energy world consumption per inhabitant ("per capita"), C_{pc}, at the time of N_{hm}, would be:

$$\mathbf{C_{pc} = C_m/N_{hm} = \left[2 \times 10^{21} \left(ZJ/year\right)/10^{10} \left(inhabitants\right)\right] = 200 \ BJ/\left(inhabitant \cdot year\right)} \tag{1}$$

where BJ = 10^9 J = billion Joules, for a hypothetical scenario which predicts equality in the future energy consumption pattern for all the world's population. It is shown in the following graphs that the developed countries are in general the largest energy consumers, and having the highest energy per capita consumption rates, just to confirm that global indexes do not represent local realities.

According to the indexes presented by Tomabechi [6], to start the projections of N_{hm}, since nuclear power accounts for approximately 5.4% of the total energy consumption in the world [1], the demand for this resource would be:

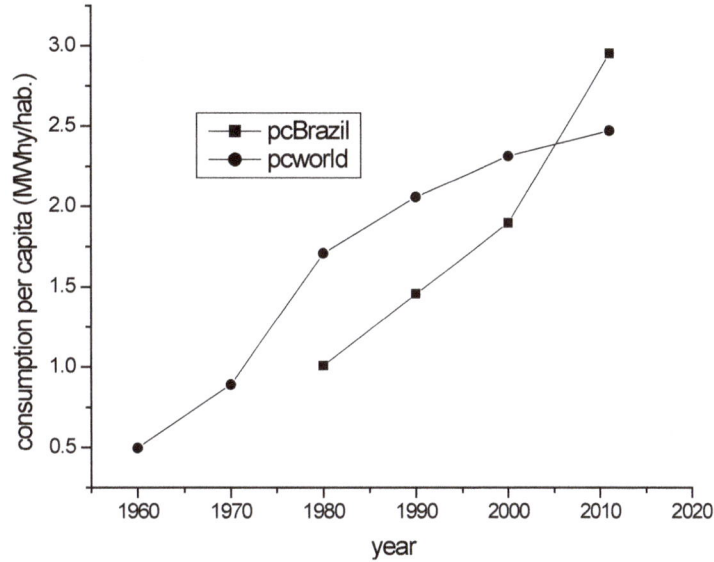

Figure 2. Consumption per capita, Brazil.

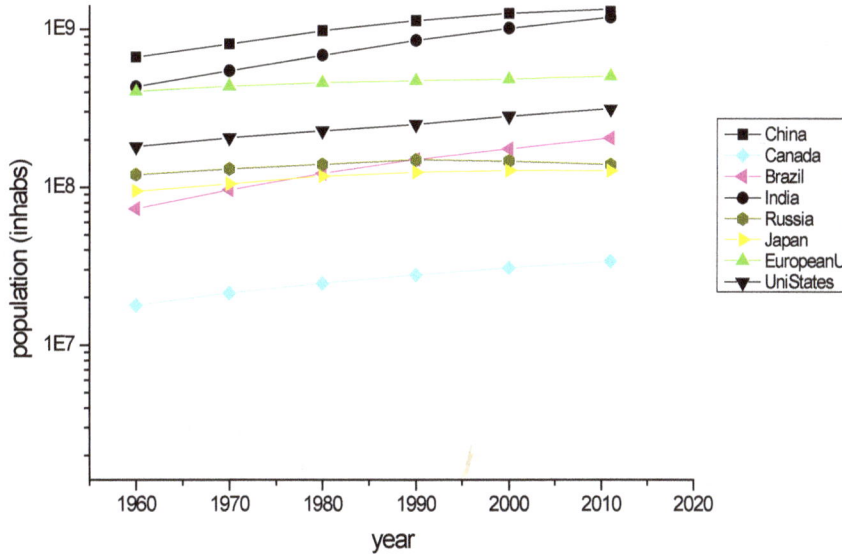

Figure 3. Population growth of the most populated countries.

$$C_{pcnuke} = 10.8 \ BJ/(\text{inhabitant year}) \tag{1'}$$

From the IBGE data [7], in 25 years, the Brazilian population N_{hBr} will reach its maximum at 219 M inhabitants (**Figure 1**, where $1M = 10^6$), at about the same time the world population is estimated to reach N_{hm}. It can be predicted that, taking into account the above hypothesis, the Brazilian energy consumption per year, C_{Br}, will be:

$$C_{Br} = Cpc \cdot N_{hBr} = 200 \times 10^9 \ (J/\text{inhabitant year}) \times 219 \times 10^6 \ (\text{inhabitant}) = \textbf{0.0438 ZJ/year} \tag{2}$$

corresponding to 2.19% of the global energy needs.

From an estimate based on the data of the 2005 Brazilian National Energy Balance presented in [15], assuming a constant growth rate of 0.67% per year in relation to the total energy consumption, and 0.64% relating to the electricity consumption, some estimates can be carried for the total energy consumption per inhabitant in Brazil, $C_{pcBrtotal}$.

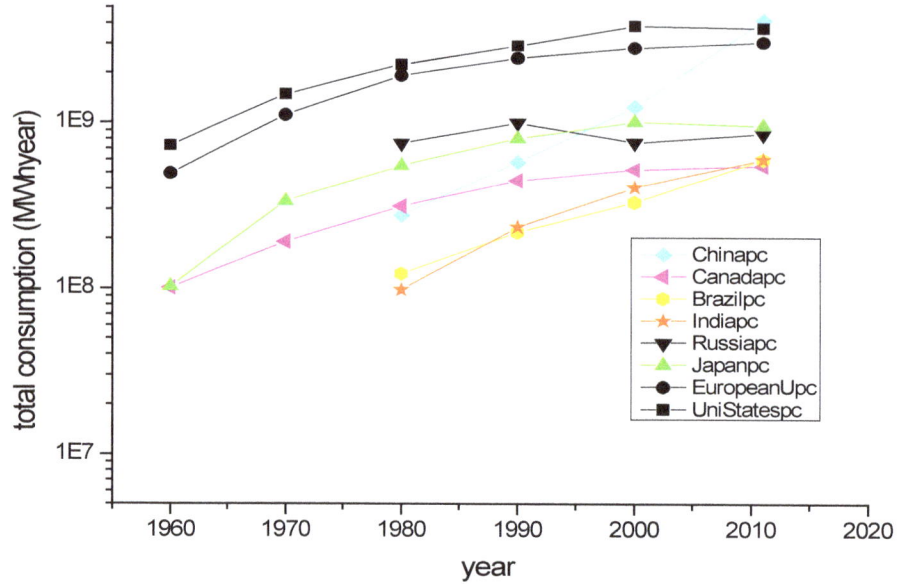

Figure 4. For the same countries of **Figure 3**, total energy consumption.

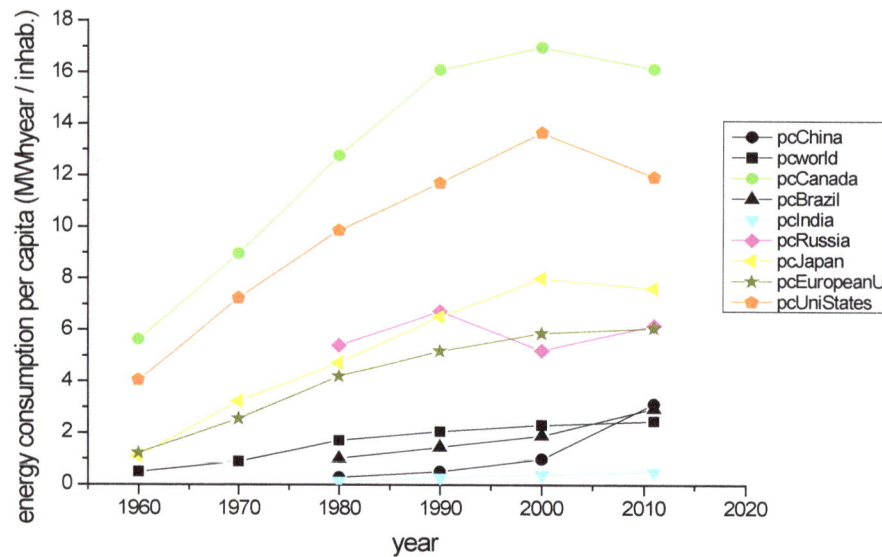

Figure 5. For the same countries, energy consumption per capita.

First, for the total energy consumption, the result is, for 2030:

$$\mathbf{C}_{\text{pcBrtotal}} = \left[9.518 \times 10^{-3} \left(\text{ZJ/year} \right) / 216{,}010{,}430 \left(\text{inhabitants} \right) \right] = 44.1 \ \text{BJ}/\left(\text{inhabitant year} \right) \tag{3}$$

and for 2039:

$$\mathbf{C}_{\text{pcBrtotal}} = \left[1.0175 \times 10^{-2} \left(\text{ZJ/year} \right) / 219{,}124{,}700 \left(\text{inhabitants} \right) \right] = 46.4 \ \text{BJ}/\left(\text{inhabitant year} \right) \tag{4}$$

both results nearly 4.5 times lower than the value of C_{pc}, predicted according to Tomabechi methodology [6], and to the relation in Equation (2). This result shows that the differences can be very high if the estimates are taken from global indexes, which are average values, justifying the use of domestic indexes for the estimates. The goal is not to criticize or favor any methodology, but just to point out that there is a need to present them comparatively, if working together on the reduction or sharing pollution/clean energies deployment costs.

Taking into account a Brazilian scenario provided by EPE/Eletronuclear [15] about the future participation of the nuclear power in the energy matrix in Brazil, and the INB [4] projections for the Brazilian uranium supplies, its future contribution to the energy demand can be estimated, and the results compared to those from Equation (2).

According to the scenario devised in the EPE document [15], nuclear energy participation in the national electricity demand would grow from the current 2.5% (from a current national total of 78 GW) to 2.7%. Thus, it will be necessary one more nuclear power plant with capacity of 100 to 300 MW and the nuclear matrix will grow from 1950 MW to a maximum of 3550 MW, considering also the construction of Angra III. This work also stated that, in a so called simple development scenario, the estimated participation of nuclear would be 4.2%, comprising the construction of one more nuclear power plant 1300 MW power, (together with Angra III) and two other nuclear power plants of 300 MW each, adding 3200 MW to the current 1950 MW, and making up a total of 5150 MW. Finally, in a so called self-sustained development scenario, more interesting to the safe future supply of energy for the next generations, an increase of 5.7% would be supported by the construction of two new nuclear power plants 1300 MW power each (together with Angra III), and four more new modular stations 300 MW power each. The electricity generation capacity via nuclear power would grow from the current 1950 MW to 6950 MW.

Thus, the participation in the energy consumption in Brazil per inhabitant in relation to nuclear energy can be estimated according to those three scenarios, and the results are shown in **Table 3**, where MJ is equivalent to millions of joules.

Using the assumptions described above for the consumption rates for the total energy and for the energy from electricity, and the values obtained by the relations (3) and (4), we can obtain the electricity consumption values projected for 2030 (16.04%) and 2039 (15.88%). Thus, for 2030, the result of the Brazilian projected electricity consumption demand, C_{pcBrel} would be:

$$C_{pcBrel} = \left[9.518 \times 10^{-3} \left(ZJ/year \right) / 216,010,430 \left(inhabitants \right) \right] \times 0.164 = 7.07 \left(BJ/inhabitant\ year \right) \quad (5)$$

and for 2039:

$$C_{pcBrel} = \left[1.0175 \times 10^{-2} \left(ZJ/year \right) / 219,124,700 \left(inhabitants \right) \right] \times 0.158 = 7.37 \left(BJ/inhabitant\ year \right) \quad (6)$$

Considering the self-sustained scenario, electricity generation via nuclear power in the years of 2030 and 2039 in Brazil would be approximately 9×10^{-5} ZJ/year, corresponding to 7.1 GW of projected power.

5. Conclusions

Taking the current indexes for the world electricity consumption into consideration, pushed up mainly by the developed and highly industrialized countries, and assuming that they are the same for the developing countries, by the time world's population reaches 10^{10} inhabitants, calculations show that the power consumption projections for Brazil, for the adopted time horizon and when using global indexes, become overestimated, when compared with the results calculated by our national indexes. According to our studies, power consumption estimates using global indexes becomes approximately 4.5 times higher than the estimates based on national indexes. Global indexes, as taken by several works and particularly the work of Tomabechi [6] analyzed here, are not reflecting the national realities in terms of energy consumption and demand, as is the case of Brazil.

Data here used for the predictions are subjected to changes and to the accuracy of the governmental agencies at the time of their publication. In the Brazilian case, the recent discovered pre-salt oil deposits offer a possibility for a different solution to a possible energy crisis, when compared, for example, to the United States. Together with China and the European Union, they are responsible for almost 65% of the world's energy consumption, having also combined factors like large population and large energy consumption per capita. Both factors lead to high levels of emissions and to the fast depletion of the current natural resources.

However, despite the great potential of the Brazilian reserves of uranium, water, oil, etc., clean energies must also be taken into consideration in the future, to meet the 2DS requirements. For a non-GHG-emission source, nuclear is the most available candidate. For Brazil, together with the renewables, the nuclear option could be a quick start to solve the problem of emissions due to deforestation, our main source of greenhouse gases.

Construction of nuclear power plants is always a subject for debates and speculation, impacting the forecasts of consumptions of uranium and thorium, by means of a correct campaign for the public acceptance. It was stated above that a way to reach some equilibrium condition could be achieved, however, if countries agree to diffuse their energy technologies for other countries, developing or under-developed, since they are at most low carbon emitters and usually still don't have enough technology development to carry out projects on clean energies. Sharing the technologies could be a fast contribution for the solution to the growing problem of climate changes.

Relating to nuclear energy, at least four factors, mentioned is this work, can put Brazil in a comfortable position in front of a possible crisis in the future demand for energy: 1) uranium: since we have enough uranium in the Brazilian territory, and counting with the future improvement of the process of uranium extraction from phosphates; 2) thorium: since Brazil has also enough reserves, for future uses in new technology nuclear power reactors; 3) population growth stability: according to the governmental projections; and 4) nuclear fuel cycle know-how.

Climate instabilities can reach us all in a near future, and global warming is one of its possible triggers. Besides, human activities play a very important role in the contribution to the instability of the weather in a global scale. The increase in the demand for energy, in the population growth, mainly in the last 50 years, leading to the increase of the greenhouse gas emissions and to all the climate instabilities we have been facing, should be assessed considering the appropriate indexes. Thus, according to this work, a different criterion must be stated, for the world to face this problem fairly, since a huge amount of investments will be needed to control or to reduce the threat of the so called "non-turning point" phenomena, in terms of the global weather.

Thus, the future transition between the massive global uses of fossil fuels to a massive global use of the alternative sources, supported by our nuclear know-how, could be carried out in such a way to help us in the improvement of the life quality in our country, and as a consequence, in our planet.

References

[1]   http://www.cia.gov/library/publications/the-world-factbook

[2]   Haratyk, G. and Forsberg, C.W. (2012) Nuclear Renewable Energy System for Hydrogen and Electricity Production. *Nuclear Technology*, **178**, 66-82.

[3]   Empresa de Pesquisa Energética—EPE (2012) Ministério de Minas e Energia, Balanço Energético Nacional 2012.

[4]   http://www.inb.gov.br

[5]   IEA (2012) Tracking Clean Energy Progress—Energy Technology Perspectives 2012 Excerpt as IEA input to the Clean Energy Ministerial. 82 p.
        http://www.iea.org/publications/freepublications/publication/Tracking_Clean_Energy_Progress.pdf

[6]   Tomabechi, K. (2010) Energy Resources in the Future. *Energies*, **3**, 686-695.

[7]   Statistical Series, Instituto Brasileiro de Geografia e Estatística (IBGE).
        http://www.ibge.gov.br/series_estatisticas/exibedados.php?idnivel=BR&idserie=POP300

[8]   Piore, A. and Planejando, C.N. (2011) Scientific American Brasil, Edição Especial Energia Nuclear, 7-11.

[9]   Cochran, T.B., Feiveson, H.A., Mian, Z., Ramana, M.V., Schneider, M. and von Hippel, F.N. (2010) It's Time to Give up on Breeder Reactors. *Bulletin of Atomic Scientists*, **66**, 50-56. http://dx.doi.org/10.2968/066003007

[10]   World Nuclear Association, WNA (2011) Current Status. http://www.world-nuclear.org/info/reactors.html

[11]   (2011) Understanding the Roles of Technology Diffusion, Intellectual Property Rights, and Sound Environmental Policy for Climate Change. World Energy Council WEC Energy Sector Environmental Innovation.

[12]   Cerri, C.C., *et al.* (2009) Brazilian Green-House Gas Emissions, the Importance of Livestock and Agriculture. *Scientia Agricola*, **66**, 831-843. http://dx.doi.org/10.1590/S0103-90162009000600017

[13]   Kidd, S. (2011) Uranium Supply for the Nuclear Future. *Energy and Environment*, **22**, 61-66.

[14]   Data to Graphics and Tables Extracted from http://www.nationmaster.com.

[15]   Geração Termonuclear (2008) Boletim da Empresa de Pesquisa Energética EPE. No. 20080512-7, Ministério de Minas e Energia, Secretaria de Planejamento Estratégico, Maio de 2008, 13.

[16]   Gonçalves, O.D. (2011) Brasil Ainda Hesita na Área Nuclear. *Scientific American Brasil, Edição Especial Energia Nuclear*, 35-39.

[17]   Ashley, V., *et al.* (2010) The Commercially Viable ADTR Power Station. *Nuclear Future*, **7**, 41-45.

[18]   www.epe.gov.br/PNE/20070625_09.pdf

[19]   Análise Retrospectiva (2008) Boletim da Empresa de Pesquisa Energética EPE. No. 20080512-1, Ministério de Minas e Energia, Secretaria de Planejamento Estratégico, Maio de 2008.

5

Transforming Energy Usage: It's Not Only about Solar

Melissa Matlock

GRID Alternatives, Riverside, USA
Email: mmatlock@gridalternatives.org

Abstract

GRID Alternatives, a non-profit solar contractor, installs solar electric systems for low-income families. Part of GRID Alternatives' program is to provide solar electric systems that are designed to replace 75% of the homeowners' electricity usage with solar power. This leaves 25% of their bill still to be paid. In order to save our resources, one must first use conservation practices, then energy efficiency, and then follow-up with renewable energy to cover the rest. GRID Alternatives Inland Empire (GRID IE) educates our participating homeowners and community members on this philosophy. However, measuring whether or not our families have been following this philosophy is hard to prove. It may seem obvious that if we want to know whether our homeowners are saving energy, we should look at their energy usage before and after solar. However, this is not the case with our low-income families that could be using electricity to make their lives more comfortable. GRID IE developed a survey to be given before homeowners received their solar systems and started their participation with GRID Alternatives and the same survey to be given after they have received their solar systems. This before and after survey (pre-test/post-test) asked our home-owners to rate their responses to 7 questions on a scale of 1 - 10. The before and after responses for each person were compared, and as a group, their differences were calculated to find out if the differences were statistically significance (within subjects, dependent Z test). 6 out of 7 questions showed statistical significance. The big picture is that change is happening among our low-income homeowners and has happened for many of the varied energy saving methods discussed. It is important to transform energy usage, because the solution is not just solved with solar.

Keywords

Energy Efficiency, Residential, Solar, Survey, Energy Conservation, Water Savings, Low-Income

1. Introduction

There are certain ways to measure energy usage. The most efficient and straight-forward method is to look at one's electric bill and gather their kW, a basic unit of power, and their kWh, the unit of energy that represents power being used over time [1].

Part of GRID Alternatives' program is to provide solar electric systems that are designed to replace 75% of the homeowners' electricity usage with solar power. This leaves 25% of their bill still to be paid. GRID Alternatives sees solar power as the solution but not the only one. Protecting earth's resources does not work with a business-as-usual approach, and cannot be saved with replacing one's usage, previously dependent on fossil fuels, with renewable energy 100% [2]. It is a more wasteful approach and just keeps the status quo. In order to save our resources, one must first use conservation practices, then energy efficiency, and then follow-up with renewable energy to cover the rest [2]. GRID Alternatives Inland Empire (GRID IE), a Southern California regional office working in San Bernardino, Riverside, and Inyo Counties, educates our participating homeowners and community members on this philosophy. However, measuring whether or not our families have been following this philosophy is hard to prove. It may seem obvious that if we want to know if our homeowners are saving, we should look at their energy usage before and after solar. However, our families are low-income and have expressed that they were not previously using their air conditioner and other appliances because they could not afford it. Since receiving solar, our families could have decided to increase their usage of these appliances to live more comfortably.

2. Methods

2.1. Survey

Here at GRID IE, we developed a method to take qualitative information from our homeowners and use statistics to create quantitative results and give GRID IE a better scope of our work and impact in energy education. We developed a survey to be given before homeowners received their solar systems and started their participation with GRID Alternatives and the same survey to be given after they have received their solar systems. This before and after survey (pre-test/post-test) asked our homeowners to rate their responses to 7 questions on a scale of 1 - 10. The before and after responses for each person were compared, and as a group, their differences were calculated to find out if the differences were statistically significance (within subjects, dependent Z test) [3]. The questions represented several principles regarding energy conservation and energy efficiency. The questions varied from easy and quick fixes to some more costly options.

2.2. Statistically Significant

For something to be statistically significant, it means that the number you get from your sample, when it is compared to a general population, is extremely unlikely to happen by chance and that it happened because of change [3]. There are levels of significance and saying that something happened 100% due to change is not going to happen. Instead, the typical levels of significance are 10%, 5%, and 1% [3]. 10% is the lowest level of significance and it means that there was a 10% probability that the number you received from your sample happened by chance. 5% is harder to prove, as it means that there is a 5% probability that the number from your sample happened by chance instead of change. At 1%, there is a 99% probability that the number did occur by change.

2.3. Hypothesis & Tailed Tests

Typically going into research, you have an idea or a hypothesis about what your results will be. When you have an idea about what the response will be, you will calculate your responses and find significance using a 1-tailed test, meaning that you know your response is either positive or negative, instead of being unsure about the direction [3]. If you just assume there will be change, but you don't know if it's positive or negative, then you use a 2-tailed test. In the research from GRID IE, only 1 of the 7 questions involved a 2-tailed test.

2.4. Sample Size

The seven questions asked to our homeowners were sent out to every GRID IE homeowner, in both English and

Spanish, over 400 families, with only 67 responses back. All 67 responses responded to both the before and the after survey.

3. Survey Responses

The survey questions are listed below with our hypothesis on what the response should be and the significance of the response.

3.1. Question 1: Energy Conservation

Q1: On a scale of 1 - 10, 1 being never and 10 being always, how often did/do you leave your lights on in a room even if you were/are not in the room?

This question is related to energy conservation. Conserving energy involves changing your behavior to save energy, like turning off your lights when you are not in the room. If our homeowners did participate in this technique, then the responses should show a decrease on this scale. Overall, the results were statistically significant at the 1-tail, 10% level. This means that there was a 10% chance that we got these answers by luck and a 90% likelihood that these answers represent change.

3.2. Question 2: Energy Conservation

Q2: On a scale of 1 - 10, 1 being never and 10 being always, how often did/do you leave your appliances (TV, coffeemaker, computers, etc.) plugged into the wall 24/7?

This question is related to energy conservation and the concept of vampire loads. If you leave an appliance plugged in, like a cell phone charger, but it is not on, it still takes energy from the outlet [4]. By educating our homeowners about vampire loads, the after responses should show a decrease in appliances being plugged in. Overall, the results were significant at the 1-tail, 10% level. This means that there was a 10% chance that we got these answers by luck and a 90% likelihood that these answers represent change.

3.3. Question 3: Energy Efficiency

Q3: On a scale of 1 - 10, 1 being none and 10 being all of them, how many of your lights were/are LED or CFLs?

This question is related to energy efficiency. Compact Fluorescent Lamps (CFLs) and Light Emitting Diodes (LEDs) use at least 75% less energy than other light bulbs [4]. However, these light bulbs have a higher upfront cost and not necessarily something cheap for our homeowners to do. However, with GRID IE's education on switching light bulbs, the responses should show an increase in these lights being used. Overall, this hypothesis showed to be true. Our results were significant at the 1-tail, 5% level. This means that there was a 5% chance that we got these answers by luck and a 95% likelihood that these answers represent change.

3.4. Question 4: Energy Efficiency

Q4: On a scale of 1 - 10, 1 being none and 10 being all of them, how many of your appliances were/are ENERGY STAR products?

This question is related to energy efficiency and is highly affected by the age of the current appliances. Appliances older than 10 years old should be replaced and appliances with the ENERGY STAR label use less energy than the other, similar products [4]. This energy saving technique is one of the more expensive methods and could be difficult for our low-income families to do. However, GRID IE has a strong focus in promoting ENERGY STAR, even developing a document about this savings technique. Therefore, the responses should show an increase in ENERGY STAR products being used. Out of all the questions, this is the question with the highest significance level. Our results were significant at the 1-tail, 1% level. This means that there was a 1% chance that we got these answers by luck and a 99% likelihood that these answers represent change.

3.5. Question 5: Air Conditioning

Q5: On a scale of 1 - 10, 1 being never and 10 being always, how often did/do you run your air conditioner during the summer?

This question was asked to supplement our general theory about our homeowners not affording their air conditioning before and now using air conditioning since they have solar. However, we did not know if this was the case, so we performed a 2-tailed test, because we do not know if this was a common theme with all of our homeowners. Our results were not significant at any level. However, the value from the calculations was positive, so it highlights that our families were using their air conditioner more, but is not statistically proven. It is enough evidence to verify our suspicions about using kWh alone to highlight energy efficiency and energy use decrease.

3.6. Question 6: Water Conservation

Q6: On a scale of 1 - 10, 1 being none and 10 being all of them, how many of your faucets were/are water conserving devices?

Recently, GRID IE has been increasing its involvement in promoting water saving as well as energy savings. They typically go hand-in-hand as it takes energy to heat and pump the water to its final location [5]. The responses should show an increase in water conserving devices, however, our results are not significant at any level. One explanation could be the old sayings "If it's not broken, don't fix it" and "Out of sight, out of mind." Faucets do not typically break and for one to transition to a water conserving device, you would have to go out and buy it. Due to water savings indirect relationship to energy conservation, this technique could be seen as having a lower return on value than switching to an ENERGY STAR product or getting new light bulbs in order to save energy.

3.7. Question 7: Irrigation

Q7: On a scale of 1 - 10, 1 being poor and 10 being excellent, how would/do you describe your irrigation watering practices (considering plant material, time of day watering lawn, etc.)?

Irrigation can be a huge waste of water and electricity. If you water your lawn after 7 am and before 7 pm, you are exposing your water to the sun, so you have a higher evaporation rate [6]. This means that you would need more water to do the same job and more energy to pump the water. Using correct irrigation practices for the season and region can help save multiple resources. GRID IE dedicates an entire slide on irrigation in the presentations we provide to our homeowners. Therefore, irrigation practices should be increasing. Overall, our results are significant at the 1-tail, 5% level. This means that there was a 5% chance that we got these answers by luck and a 95% likelihood that these answers represent change.

3.8. Table 1

Table 1 gives a quick summarization of the survey responses for the 7 questions.

Table 1. Results of the seven survey questions presented to GRID Alternatives inland empire families.

Questions	1	2	3	4	5	6	7
What was it about?	Light usage	Appliances plugged in	E.E. lights	Energy star products	Air conditioner	E.E faucets	Irrigation
What we expected?	Decrease	Decrease	Increase	Increase	Unknown	Increase	Increase
Did it happen?	Yes	Yes	Yes	Yes	It increased	Yes	Yes
Was it Significant?	Yes	Yes	Yes	Yes	No	No	Yes
If so, at what level?	10%	10%	5%	1%	N/A	N/A	5%
If so, at what tail?	1-tail	1-tail	1-tail	1-tail	Tested at a 2-tail	1-tail	1-tail

4. Conclusion

Although change has been happening, the analysis is limited. It is important to remember that the results represent significant change, but it does not show that GRID IE is the reason for it. There are many other potential reasons for the change, such as, education efforts from other organizations, discounts on prices for appliances, etc. This is a limitation to GRID IE's energy research. Due to time and resource constraints, another limitation in this research is that the pre- and post-test were delivered to our homeowners at the same time. However, the questions were designed to not lead our homeowner to feel like they need to choose higher on the scale. This did not occur on several questions, like question 6, where the families did not change their behavior and showed this by marking the same answer in both tests. The future scope of GRID IE's research would like to provide the pre-test to our homeowners starting their application then provide them the post-test during their after solar warranty training, a time difference on average of 9 months. However, in the meantime, the big picture is that change is happening among our low-income homeowners and has happened for many of the varied energy saving methods discussed. It is important to transform energy usage, because the solution is not just solved with solar.

References

[1] (2014) KW and KWh Explained. *Understand & Convert between Power and Energy*.

[2] Krieger, S. (2009) Before Adding, Try Reducing. *The Wall Street Journal*.

[3] Noviello, N. (2008) Secrets of Statistics. 9th Edition, Wiley Custom Services.

[4] Miller, P. (2009) Energy Conservation. *National Geographic Magazine*.

[5] Gies, E. (2010) Water Conservation to Save Energy. *The New York Times*, 17 May.

[6] Smith, W.B. (2008) Landscape Irrigation Management Part 5: Irrigation Time of Day. HGIC 1804: Extension: Clemson University, South Carolina. Clemson Cooperative Extension.

6

The Risks of Financing Energy in Turkey: Heading for a Rocky Road

Esin Okay

Department of Banking and Finance, Faculty of Commercial Sciences, Istanbul Commerce University, Istanbul, Turkey
Email: eokay@iticu.edu.tr

Abstract

After the implementation of Energy Efficiency Law (EEL) in Turkey, Turkish engineering firms show considerable interest in energy savings projects. There is a growing sign of an energy efficiency implementation in Turkey. However, energy importation costs of Turkey still present a challenge leading to huge financial burdens. This paper explores the energy financing in Turkey and the likelihood of risks after the establishment of structural reforms. The so-called structural problems on the agenda of Turkey are emphasized during the crisis. Actually, global crisis affected the lending conditions herein counterparty risk and restricted the relations of enterprises and banks. In this study, it is argued that there are strong barriers for the future energy saving industry in Turkey made up of macro-economic, micro-economic and financial risks coming through both domestic and global means.

Keywords

Energy Projects; Energy Saving Industry; Energy Financing; Credit Risk; Country Risk; Turkey

1. Introduction

There has been a great interest in energy efficiency improvement since the first oil price shock in the early seventies, and recently interest has heightened further because of the global warming effects of high energy use. This three decade long experience in implementing energy efficiency projects in the OECD countries has provided substantial documentation of both the economic and the environmental benefits of adopting energy efficiency improvement measures and policies. Yet, even in these developed economies, there remain a number of barriers to more widespread application of energy efficiency measures [1].

After the first oil crisis1973, energy service companies emerged in the United States. An energy service company is an enterprise that fully provides integrated energy services to their customers (mainly large energy users, but also utilities), which may include implementing energy-efficiency projects (and also renewable energy projects), frequently on a turn-key basis [2]. It provides performance and savings guarantees, and its remuneration is directly tied to the energy savings achieved. Therefore, an energy service company risks its payments on the performance of equipment and services implemented. Some of them finance projects, recovering their investment cost from the resulting savings [3].

The energy consulting concept has been established first in North America at the very beginning of 80's (see authors [1] [4] [5] who reviews the US energy saving industry and the concept of energy efficiency projects). Then, the concept spreads to Europe where the energy saving industry has successfully developed in some countries, such as Germany, but not others in EU. Only after the 90s, the energy saving concept were created in developing countries. At that time, energy end-users have just went through the first energy crises and they were looking for important operation cost reductions. Many major energy customers were developing a new way to manage and monitor their energy consumption, looking to install the most efficient equipment to replace old inefficient ones. The needs for third party financing mechanism or energy service performance contracting [6] had been proposed by some engineering companies already involved in energy management, since it was perceived as one of the main barriers to implement those potential projects. So for more than 10 years, energy saving projects operated in North America and then, in a lesser way, in Europe, based mainly on the economic benefit of the proposed projects [7].

Nowadays, international authorities are looking for solutions for the increasing energy needs and environmental impact of energy consumption and energy services. World Bank has developed over three dozen projects to support the development of local energy saving industries, including Brazil, Bulgaria, China, Croatia, India, Poland, Thailand, Tunisia, Turkey, Uruguay and Vietnam. While the development of local energy saving industries can often take more than a decade, the benefits can be substantial [8]. Bertoldi, Boza-Kiss and Rezessy (2007) give an update of the status report of energy savings in Europe to investigate the specific situation in EU-27 and in some non-EU countries. The authors find the energy markets in Europe to be at diverse stages of development (Germany and Italy: large numbers of energy projects, Latvia, Romania, and Denmark: a few energy companies, Albania, Serbia, Hungary: decreasing energy market, Estonia, Greece, Belarus, and Macedonia: just getting established, Italy and France: expanding [9]. Turkey, under the regulations of EU directives, is another country that established a new energy saving industry in EU [10] [11].

Turkey concentrated on the development of energy management by General Directorate of Renewable Energy of Ministry of Energy and Natural Resources in cooperation with international organizations such as WB, EU and Japan International Cooperation Agency (JICA) since 1995 [12]. Energy industry in Turkey is endowed with rich renewable energy sources (hydro, wind, geothermal, and solar power) potential. The technically feasible potential of hydro amounts to 216 TWh/year while economically feasible potential is nearly 125 TWh/year. About 32% of the economically feasible hydro-potential has been developed, based on average annual generation [13].

After the adoption of the Energy Efficiency Law (EEL) 2007, Turkey started to conduct energy efficiency (EE) projects regarding the precautions to increase the efficiency of energy consumption by industrial establishments. There is a rapid change on the way to building up energy saving industry in Turkey. Currently there are 34 certified energy companies in Turkey. The market is made up of local capital owners with no joint venture or foreign energy company. Most of the energy projects are connected with electricity rather than oil. Despite some teething problems relating to financial support, the energy saving industry in Turkey is slowly but steadily growing. The industry conceive to finance energy-efficiency projects with the help of European Union (EU) and World Bank (WB) funds in the short run and with Turkish banks' credits in the medium to long run. For instance, the government is cooperating with the WB for supporting small hydro and wind energy projects [12].

Turkey's energy demand has been growing with a rate of 6% for decades and this demand is expected to persist as a result of rapid urbanization and industrialization. Primary energy demand has been projected to reach 220 million toe in 2020 (a 150% increase compared to the current level). The limited availability and production capacity of domestic energy sources cause import dependency primarily on oil and gas [10]. WB has allocated 420 million US dollars credit to Turkey for supporting developments in the utilization of renewable energy sources. Two development and investment banks, in particular the Turkish Industrial Development Bank (TSKB) (75% of the credit) will act as intermediary in the use of the credit most of which will be used for development of geothermal, hydro and wind energy sources. Only private sector institutions will be able to utilize the credit. The project proposals should be in conformity with the Turkish environmental legislation as well as other conditions which could be requested by the WB [14].

The organization of this paper is as follows: Section 2 briefly reviews energy saving projects and energy financing mechanisms. In Section 3, the counterparty risk and credit risk via energy financing have been presented, emphasizing the experiences in the developed and developing countries. Section 4 explores the financing of energy projects in Turkey, supporting with data of bank credits. Lastly within Section 5, the future composition and likelihood risks of energy industry in Turkey are revealed.

2. Energy Savings and Project Financing

The financing mechanism of energy projects is generally classified as the "guaranteed savings" and "shared savings". In the "guaranteed savings" mechanism, the energy company guarantees a certain level of energy, savings sufficient to cover clients' annual debt obligation, and protect the client from any performance risk, and the clients are financed directly by banks or by a financing agency. The client repays the loan and the credit risk stays with the lender. In countries with established banking structure, project financing, and stable economy the "guaranteed savings" mechanism functions properly [10]. Success stories are available for the UK, Austria, and Hungary [5] [15] [16].

In "shared savings" mechanism, energy company carries both the performance and credit risk. Energy company repays the loan and the credit risk stays with the company and the client assumes no financial risk. In developing countries, the "shared savings" mechanism is more suitable since it does not require clients to assume investment-repayment risk. The client assumes no financial obligation other than to pay a percentage of the actual savings to the energy company over a specified period of time. This obligation is not considered debt and does not appear on the customer's balance sheet. In such markets, there are too few energy service companies most of them small-sized. There are WB funded stories available from India, China, and Brazil [1] [17] [18]. The professionals involved with energy efficiency projects have long believed that the lack of financing mechanisms was one of the main barriers to the establishment of the industry in Brazil [18]. Most Mexican energy companies are very small engineering firms that lack capital to invest in projects. To overcome this financial barrier inte- raction with international energy companies and to seek possible business partnerships is admitted [19].

Financial barriers are found to be the lack of funds for awareness and the insufficiency of financial support of energy companies. Bannai *et al.* (2007) using financial derivatives and actual data from existing plants, show how the concepts and tools of financial engineering can be used to hedge the risks due to volatility of fuel and electricity costs to increase the stability of the profit associated with energy business [20].

The financing options available for energy performance contract projects are bank financing, direct customer financing, public financing (bonds), energy service companies or third party financing [10]. Whereas, the policy solutions to promote project financing need developing an International Energy Efficiency Financing Protocol (IEEFP) and creating an Energy Efficiency Projects (EEP) Financing Fund [21]. Bank financing of energy companies, instead of municipalities, is a well-accepted model which allows the entry of private capital into the sector and offers instant modernization projects. This is at minimal or no increase in costs to the municipality as energy companies receive service fees from municipalities based on energy savings achieved [22]. Similarly, The European Bank for Reconstruction and Development (EBRD) is committed to developing more efficient uses of energy within its countries of operations. One way of promoting energy efficiency improvements is through energy service companies [23].

The need for increase efficiency and environmental benefit is today even stronger in developing countries than it is in industrialized ones. Brazil, India, China [24] are probably the three best examples of those needs, and on the potential global environmental impact if these goals are not achieved. In this context, the need to investigate what are the actual barriers for the full development of this concept in countries like India is quite important [25]. The energy saving industry in India is still under evolution. Energy companies in India face a number of surmountable barriers that need to be worked on. Most of the barriers will fall with the lapse of time as the market becomes more familiar with the business, but some do need action by large organization or the government. The concept is new and not widely known. Customers are reluctant to sign long-term contracts. There are a few energy companies, most of them small-sized, in India [1]. WB is funding a project on developing financial intermediary mechanisms for energy-efficiency investments in some countries like India. Surveys have been conducted on Indian energy saving industry, energy audit and consulting companies regarding their perception of energy companies and energy projects in India [26].

On the other hand, Hungary has been a leading model to develop the scope of energy savings as one of the most successful developing country in Europe. With a strong third party financing market, early restructuring energy sector, good institutional and banking sector reforms Hungary can be shown as a good indicator for positive results [27].

Hungary pioneered with the market-oriented restructuring of the economy, starting already in the 1980s. Within the economic reforms, it has been a leader in electricity restructuring in the region by completing a large share of unbundling, price liberalization, privatization, lifting of most subsidies, and discontinuing cross-subsidies by the late 1990s. By 2004, all non-residential electricity consumers are eligible to choose their supplier; and the gas market has been opened as well. There are about 10 - 20 key players and some other 200 more players acting in the Hungarian market. The ownership of the district heating systems was transferred to the local governments, heating stations were privatized, direct subsidies were stopped and also state owned flats were privatized. Considerable amendments have been made in the legal framework as well. Devolving statutory powers to local authorities, fiscal decentralization and rules of public budgeting related to energy savings have also greatly contributed to the development of energy saving projects in the municipal sector. Last, but not least, payment arrears and commercial losses are much less of a problem in Hungary than in the rest of the region [15].

After the change in regime, Hungary has been a prime target for various international donor and aid programs targeting energy efficiency. Many international actors, including the EBRD, the Global Environment Facility (GEF) through the International Finance Corporation (IFC), and the United State Agency for International Development (USAID), have been supportive in the promotion of energy saving industry in particular through different approaches and programs [15]. IFC addresses Hungary as an innovative model for energy market experience with decreased operating costs for companies and hence increased international competitiveness [28].

Some other factors contributed to the Hungarian success in the development of a strong energy efficiency. The requirements for EU accession meant providing a strong imperative to liberalize the energy sectors and ensure that energy prices reflect true economic costs. The requirement for lower energy intensity also exerted some pressure to trigger institutional changes in relation to the improvement of energy efficiency in the country. More recently, the ratification of the Kyoto Protocol by the EU and the accessing countries (including Hungary), forced countries to introduce policies and measures to promote reductions in greenhouse gas (GHG) emissions, which generally implies the conservation of energy, since energy consumption is by far the biggest contributor to greenhouse gas emissions and improving energy efficiency among the most cost-efficient options for GHG reduction. These measures contributed as well to the development of stronger incentives to develop the energy efficiency sector [15].

3. Financing Risks of Energy Projects

Economic developments after 1980's has led to an increased liberalization of markets and countries have to find other strategies to ensure the proper management of their projects and investments. Nowadays, there is a huge potential of risks through liberalization endangering the economic and financial system. Therefore, sectors in economies do vary within the scope of this new political order, so as the energy sector's fall [6].

Energy companies have strong transition impact as a private sector instrument to deliver energy savings. In this respect, the role of enabling policies should be noted in national energy policy act, state regulations for performance contracting and utility programs. These companies may finance or assist in arranging financing for the operation of an energy system by providing a savings guarantee. Energy service companies operate under an energy performance contracting arrangement, whereby they implement a project to deliver energy efficiency, or a renewable energy project, and use the stream of income from the cost savings, or the renewable energy produced, to repay the costs of the project [23].

Energy savings performance contract is designed to establish energy efficiency projects. Third party financing is the most commonly used type for energy projects and it can be realized by a guaranteed saving contract. Country risk, credit risk and counterparty risk play a major role via projects with too high level of risk as for the case with energy projects. So, energy financers should have strong balance sheets. They must have strong assets to take on huge liabilities of clients that have long-term projects. This is extremely a big risk that points to a considerably significant size-based balance sheet to finance projects. Therefore, energy companies should be committed to risk management, as well. Energy projects should be managed with risk reduction methods like hedging instruments and venture capital. Therefore, when energy companies need financial support to invest such risky projects banks analyze detailed financial statements before extending credit. Besides, banks should implement BASEL II, and then BASEL III. The purpose of BASEL II-III risk implementations is to create an international standard that banking regulators can use when creating regulations about how much capital banks need to put aside to guard against the types of financial and operational risks banks face. BASEL III is a response to prevent the weaknesses in the BASEL II framework where it will lead tightening the risk management approach. BASEL III is expected to be implemented globally between 2013 and 2019 [29]. BASEL III could tighten banks' provision of long-term finance or result in increased interest rates. Either of these changes and tightened capital and liquidity requirements are likely to limit the amount of capital available for renewable energy financing from banks in the future. Together, these are threats to renewable energy deployment because limited financing may prevent the energy projects [30].

Similarly, energy companies should anticipate macroeconomic factors of the country, foreseeing the risks [21]. Country risk covers credit risk, as both refers to investing in a country. For example, the Country Risk Assessment Model (CRAM) of OECD produces a quantitative assessment of country credit risk based on three groups of risk indicators (the payment experience of the participants, the financial situation and the economic situation) [31]. Credit risk is the risk of loss due to default by a counterparty on its contractual obligations, or as a result of reduction in the credit quality of the counterparty [32]. So, in case of energy investment projects with high risk credits the country risk is highlighted. Counterparty credit risk is the risk that the counterparty to a financial contract will default prior to the expiration of the contract and will not make all the payments required by the contract. Only the contracts privately negotiated between counterparties—over-the-counter derivatives and security financing transactions—are subject to counterparty risk. Exchange-traded derivatives are not affected by counterparty risk, because the exchange guarantees the cash flows promised by the derivative to counterparties. Counterparty risk is similar to other forms of credit risk in that the cause of economic loss is obligor's default. There are, however, two features that set counterparty risk apart from more traditional forms of credit risk: The uncertainty of exposure and bilateral nature of credit risk [33].

Third-party finance (banks or special funds) can be used to provide customers with the required funds to pay for these projects, isolating the energy company from credit risk as depicted in **Figure 1** on the part of the customer. The guaranteed savings scheme is likely to function properly only in countries with an established banking structure, high degree of familiarity with project financing and sufficient technical expertise, also within the banking sector, to understand energy-efficiency projects (e.g. the UK, Austria, and more recently, Hungary) [16]. Financing may be provided by a separate financial sub-division of the energy industry for larger projects with

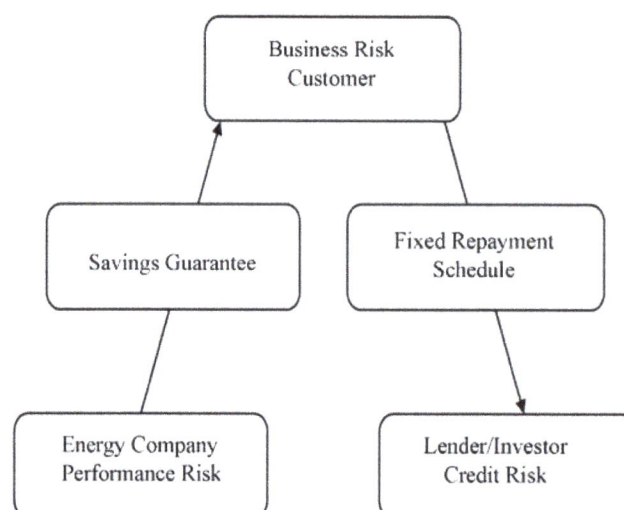

Figure 1. Guaranteed Savings (GS) financing mechanism of performance contracting in energy projects [35].

either good default options or external guarantees provided by the government or trustworthy international lenders [34].

The interest and co-operation of financial institutions and banks are essential for energy market development financing of individual projects. Government and donor agencies can stimulate the market for affordable financial options through various means: Soft loans or grants for projects and energy business expansion, support to demonstration projects for information dissemination, information and education programs for financiers to ensure they understand the financial soundness of energy efficiency projects, evolving appropriate financial guarantee mechanism against loans [21].

Energy services projects, in all their varying forms, are technically complex undertakings, requiring significant design, engineering and development efforts to create a transaction that is economically compelling to the end user customer. However, all these efforts could be for naught if third party financing is required and the transaction is not structured to satisfy the requirements of the capital markets. Developing a bankable energy services transaction requires a thorough understanding and balancing of the needs and objectives of the three main parties to the deal-the customer (end user or obligor), the energy company, and the lender. Obligor credit risk is the core risk for all financing transactions and simply translates to the end user's ability to satisfy all of its obligations (payment and otherwise) under the terms of the transaction. Despite the wide variety of financing structures for energy services transactions, their common foundation is the end user's obligation to make payments in exchange for the products or services received. In whatever forms it takes (*i.e.* lease payment, loan payment, services payment, usage payment, etc.), that payment obligation provides the lender's primary source of debt service and return on the equity investment, if any. As a result, the bulk of the lender's underwriting activities are focused on a detailed analysis of the end-user's overall credit profile. Operating performance, cash flow, debt service capacity, liquidity, balance sheet strength, market position, management capabilities, and future projections are among the many factors thoroughly examined during the underwriting process. In order to achieve the desired end result, i.e. transaction approval, the credit analysis must conclude that the obligor is capable of fully satisfying its obligations throughout the full term of the transaction [36].

Most facility managers agree that energy management projects are good investments. Generally, energy projects reduce operational costs, have a low risk/reward ratio, usually improve productivity, and even have been shown to improve a firm's stock price. Despite these benefits, many cost-effective energy projects are not implemented due to financial constraints. A study of manufacturing facilities revealed that first-cost and capital constraints represented over 35% of the reasons cost-effective energy projects were not implemented. Often, the facility manager does not have enough cash to allocate funding or cannot get budget approval to cover initial costs. Financial arrangements are one way to mitigate risks of a facility's funding constraints, allowing additional energy savings to be reaped.

Loans have been the traditional financial arrangement for many types of equipment purchases. A bank's willingness to loan depends on the borrower's financial health, experience in energy management, and number of years in business. Obtaining a bank loan can be difficult if the loan officer is unfamiliar with energy projects. Loan officers and financiers may not understand energy-related terminology. In addition, facility managers may not be comfortable with the financer's language. Thus, to save time, a bank that can understand energy projects should be chosen. Most banks will require a down payment and collateral to secure a loan. However, securing assets can be difficult with energy projects, because the equipment often becomes part of the real estate of the plant. For example, it would be very difficult for a bank to repossess lighting fixtures from a retrofit. In these scenarios, lenders may be willing to secure other assets as collateral [37].

4. Energy Financing in Turkey

Turkey's energy market is sponsored by government and WB funds during its incubation stage under the shared savings mechanism (SS) is depicted in **Figure 2**. As stated before, under a shared savings the energy company assumes both performance and credit risk (as the client takes over some performance risk, it will try to avoid assuming any credit risk) [16] [21].

Creating a competitive, mature energy market like Hungary (as a developing country model, has more than 200 energy service companies) through the GS mechanism as depicted in **Figure 1**, is not possible under today's Turkish economic environment. The SS concept is a good introductory model in developing markets because customers assume no financial risk, which they like or even demand. The downturn of this structure is that it

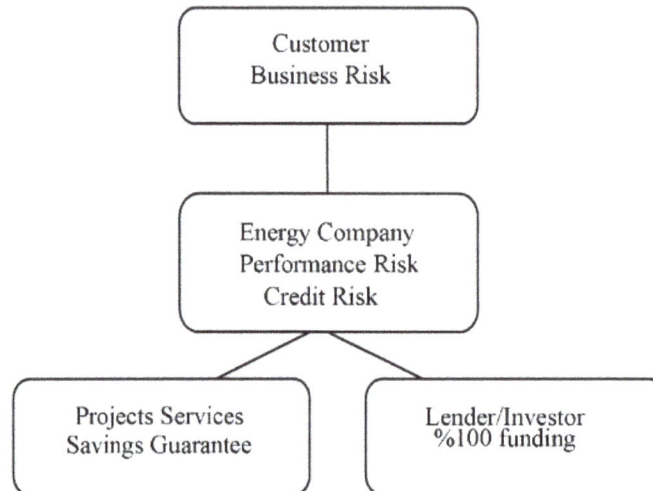

Figure 2. Shared Savings (SS) financing mechanism of performance contracting in energy projects [35].

limits long-term market growth and competition between energy companies and between financing institutions because small and/or new companies with no previous experience in borrowing and few own resources are unlikely to enter the market if such agreements dominate [38].

Nowadays, there are 13 domestic banks in Turkey proposing credits to enter the energy efficiency market. Most of them are commercial and the rest are state banks [12]. The energy projects and the credits financing in Turkey had a volatile trend throughout the decade coinciding with two crises (domestic crisis; 2000-2001 banking crisis and 2008 global crisis). The government unfolded privatization projects in early 2000's, primarily the energy projects to fulfill household requirements like natural gas contracts and electricity contracts.

Therefore, from late 2000 to early 2004, the share of energy credits rapidly went up high reaching approximately up to 14% around 2002-2003 as depicted in **Figure 3**. The share of energy credits are calculated from the set of data obtained from Central Bank of the Republic of Turkey, concerning years (2000-2013). Relatively, weak energy credit flows have seemingly been constrained by the 2000-2001 banking crisis in Turkey. After the crisis, the flows recovered a bit but falling gradually under 10% by 2004 energy credits continued to diminish until 2008. Later, when the implementation of EEL enhanced the energy projects in Turkey, the share of credits started to rise again up to about 7%, but not so compromising till 2013 as depicted in **Figure 3**, in the shadow of the ongoing global crisis and majorly country risk that will be discussed in the next section. The current situation of energy saving industry shows a low profile performance and a considerable credit risk due to micro-economic and financial situation in Turkey. Eventually, this points to future risks of energy industry in Turkey.

5. The Present and Future Risks in Turkey

First of all, the energy market in Turkey is exposed to macro-economic problems (country risk). Country's current low interest rate and low inflation potential due to small debt-to-GNP ratio unlocks the potential for growth in this sector. However, large debt problems of the developed countries and a worldwide fear of risk may block the foreign direct investment (FDI) flow that is necessary for the progress of the market as well for the reduction of the country's current account deficit. Turkey has persistent and risky current account deficits for years. This risk as a destabilizing factor for Turkish Economy has been studied in many papers. Turkey is exposed to macroeconomic instability, financial fragility and crisis in the case of internal and external economic and political shocks, because of the structural economic problems—low productivity and global competitiveness level in high-technology products, foreign dependency in energy, intermediate goods and finance, domestic saving gap-resulting in unsustainable current account deficit [40]-[42].

After the 2008-2009 global financial crisis, the chronic problem of the Turkish Economy, the current account balance trouble recrudesced. In the present instance, the factors increasing the current account balance are thought not only structural problems but also economic conditions, booming international credit inflows from global markets appreciating the Turkish Lira, deteriorating the current account balance steeply [43]. FDI neces-

ENERGY CREDITS %

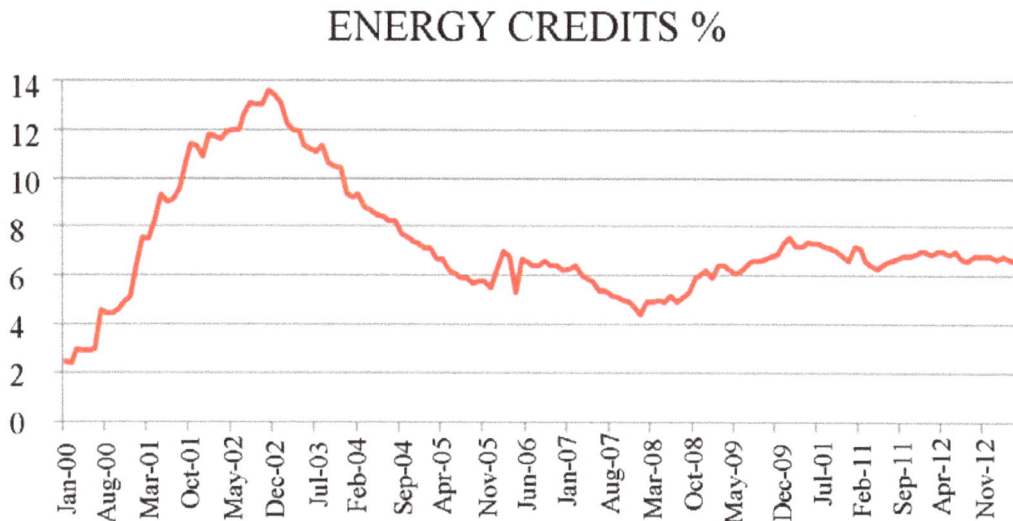

Figure 3. Energy Credits (percentage share %) [39].

sity is highlighted as the current account deficits enhanced after the crisis.

Soytas and Sari (2009) mention that Turkey is facing an investment problem and regardless of which alternative energy sources she wants to develop or utilize a large portion of this investment would be through accumulating capital based on imported technology [44]. The role of Turkey's energy market and its relationships with the universities and research & development centers had been given by [10]. It is stated that the Turkish government must support and develop a shared vision of energy-related innovation between industry, universities, government-based research & development centers, energy companies, and related (FDI). Turkey's problem with FDI is that current account deficit (CAD), in which energy imports contribute significantly, is large and getting larger (Central Bank of the Republic of Turkey, CBRT, http://evds.tcmb.gov.tr); restricting the motivation of long-term FDI. Realization of ES potential via energy-related innovations should help reduce this portion of the CAD, and in turn, should encourage FDI flow to the country. Reduction in the CAD via successful energy-related innovations is expected to create an induced positive recurring effect on FDI flow. It is very important not only, to design efficient exchange rate policy in connection with the monetary, the fiscal and the external finance policy but also, to improve global competitiveness, credibility and investment climate in order to achieve sustainable current account balance and economic growth in the long run for the Turkish Economy. Otherwise, under the challenging conditions of global competitiveness, the Turkish Economy may suffer from unsustainable high current account deficits in the future [43].

Especially, to maintain a lower credit risk of financing energy market, it is a must to align with international energy companies and to seek possible business partnerships, securing third party financing from international finance organizations (beyond WB) such as IFC, Environmental Enterprises Assistance Fund (EEAF) and the Renewable Energy Equity Fund (REEF).

Despite the banking sector's high liquidity and profit after the adoption of BASEL II, lending volumes remain depressed. This is primarily because of the risk-averse approach of the new banking system and contrarily much attributed private sector and its capital inadequacy. Unfortunately, BASEL II was not implemented by the private sector in Turkey. The unsuccessful implementations of the BASEL II for the small and medium sized firms in Turkey show that companies are not apt to implement risk management.

Turkish private sector-consisting of 99% small and medium sized (SMEs)-maintains the effects of weaknesses and inadequacies for years that provide insight on the limits to firm growth and sustainable investment. By all means, SMEs, being small in size and having structural problems, cannot bear the entire financial risk of energy projects. In Turkey, government and KOSGEB collectively provide tax incentives, subsidies and lending programs for SMEs [45]. But because companies are not apt to adopt the generally accepted rules of management the problems still continue. Therefore, only large-sized firms are on the platform to establish an energy market. If they come together to form joint ventures with domestic firms or established foreign energy companies the result might change. If not, then energy financers should have strong balance sheets, as stated before.

Energy companies must have strong assets to take on huge liabilities of clients that have projects of 7 - 10 years. This is an extremely a big risk that points to a considerably significant size-based balance sheet to finance projects. Therefore, they should be committed to risk management. Energy activities should be managed with risk-reducing methods like hedging instruments and venture capital. Also, for secured loans, it is important to assess the end-user payment default risk, energy company non-performance risk and bank's loan-repayment default risk. These risks are to be appropriately addressed in order to ensure successful projects.

However, companies in Turkey do not use risk management tools effectively. Without risk management, SMEs in Turkey especially have to bear size risk, credit risk, capital inadequacy, regulation risk, technology risk and country risk. They have problems of coping with the generally accepted accounting and financial standards that especially lead to poor management of assets and liabilities. With the lack of SMEs—that constitutes 99% percent of private sector—the Turkish energy market is inevitably is small. In such a market, candidate energy companies in Turkey—with poor assessment of risk culture—will not be ready for GS financing mechanism. After all, with a finalized adoption of BASEL II in Turkey, banks may resist financing energy projects because of the risky nature of this business. What's more, as the arrival of BASEL III is still unknown in private sector, the new accord could again tighten banks' provision of long-term finance [30]. While this is a coming threat to energy markets all over the world, it could ruin the progress of Turkish energy savings.

Similarly, another concern of SMEs will be shaping the future of energy market and energy efficiency projects. That is, an update in legal regime and norms should affect energy partners and projects which will in turn help in cooperation with foreign capital and companies to compete. In 2012, a new legislation was prepared in accordance with current legal and regulatory requirements. This legislation, called as new Turkish Commercial Code No.6102 [46], aims to force the companies to follow the generally accepted standards of accounting and auditing, transparency, eventually increasing level and change the investment environment for good. It will help to increase corporate governance and transparency level for better investment conditions. Although, a new regulation is issued in July 2012 for implementing accounting standards (in line with the international standards) to evaluate the Turkish companies, there will surely be a need of time to adapt to this new situation. The companies will be subject to the new Turkish Accounting Standards in line with the International Financial Reporting Standards on 1 January 2013 [46]. Until then up to the actual deadline for adapting given by the government, July 2014, there will still be poor measurement and inadequate accounting system for companies in Turkey, which in the end apply for the energy companies. This seems to be a long period after July 2014. The concerns will continue as poor measurement and auditing problems lead to transparency inadequacy, lack of creditworthiness and corporate governance which mean risky for prospective partners—mostly foreign ones—of energy companies (credit risk and regulation risk).

Certainly, the updated legal formalities are expected to provide a more predictable and well-designed system with stimulating policies in the near future. The modern legislative change will welcome foreign capital to the new market just like the case for the Turkish Financial Markets. At least, the target of the government is to take a path leading a similar result. The Turkish banking and mostly the Turkish insurance market happened to be the subject of the foreign capital lately. After the legislative transformation, in a few years of time, foreign capital owners rapidly invested in the Turkish insurance companies. Today, the percentage of foreign capital in the Turkish insurance sector is nearly 90%. But it should be noted that aside from the private sector enterprises, financial companies were subject to a complete transformation in corporate governance, accounting standards, risk management regulations. The rapid attraction of foreign capital to financial companies is explicable because of the highly standard financial system with modern legislations. Nevertheless, the case of financial system and efforts of financial institutions stand far beyond the SMEs in Turkey. Actually, the composition of restructuring fully needs both sides the mutually related financial sector and private sector at the same time.

Beyond such financial and risk-related issues there are not any technical-/engineering-related barriers since many Turkish private companies have already proved their qualifications by completing very successfully especially the large-scale housing as well as industrial construction projects abroad. On the other hand, there is no opportunity for the energy companies in street-lighting or similar tasks as such projects are handled by the municipalities in Turkey. So, energy companies should engage in energy-related research and development and innovations. Literature reveals that energy/emission improvements of countries may be related to their innovation and research and development-activity levels [47].

Despite ongoing changes, there is still a lack of knowledge on energy market and its facilities. Therefore, the government must enforce energy campaigns likewise incentives and investment support programs strengthened with universities and technical schools. With a lack of information and campaigns, it will be hard to erase low awareness. Especially all of the positive legislative change will have no valueable effect on the development of the market. Among all the state universities of Turkey, only Gazi University applied for the training tasks and was approved by the EIE in February 2011. There is no information on whether any other universities had applied to become energy-related centers. It is assumed that the number of such universities will not exceed three to five in the near future [11].

In the light of the barriers, the situation shortly points to an inevitable halt in progress of the energy market in Turkey. Large current account deficits, uncompleted structural reforms, unfinished privatizations make the creation of a mature, perfectly competitive energy market suspicious. It's clear that no progress could be obtained until Turkey provides a further modernization of law system adopting legal reforms and gains improved results.

6. Conclusions and Recommendations

The recent Energy Efficiency Law (EEL) has an important potential to further develop renewable energy in Turkey. But there are strong barriers for the future energy industry in Turkey. All the assessment of domestic risks including credit risk, country risk, financing risk and regulation risk show that Turkey is not fully ready for building an energy saving marketing. The share of energy sector credits shows a low profile due to micro-economic and financial risks in Turkey. Eventually, this points to future risks of financing energy projects in Turkey.

Turkey must direct its limited energy-policy related funds primarily to the private sector exhibiting high energy consumption, high energy efficiency potential, and high competence in research & development and innovation activities. Innovation flourishes as a consequence of knowledge accumulation and research & development experience, all of which also expedite compliance with laws and regulations. The new efficiency law of Turkey offers many opportunities and solutions to problems of energy savings industry:

1. for foreign energy companies to enter the new Turkish energy market;
2. for domestic energy companies to increase business volume with industrial, electricity, transportation, and construction sectors and to undertake joint ventures with foreign energy companies;
3. for Turkish universities in establishing new or developing the existing energy and energy-efficiency related institutes, departments, degree-granting programs, and EEL-related "official certificate" programs;
4. for Turkish universities in acquiring commercial and state funds to support academic research projects and faculty members, and in establishing new or raising the existing collaborations with the foreign universities in energy-efficiency related academic areas; and
5. for "domestic and foreign banks" and "credit companies" in the long-and medium-run financing of Turkish companies in their energy-efficiency related projects entailed by the EEL.

Turkish economy recovered quickly from the precipitous drop in output triggered by the 2008 global crisis. However, the crisis shortening of liquidity changed the conditions of economic units, tightened lending and led to shift in investment factors. Despite the banking sector's high liquidity and profit, lending volumes remain depressed. This is primarily because of the risk-averse approach of the new banking system and contrarily the attributed but so unfruitful private sector and its capital inadequacy. Turkish private sector (merely the SMEs)—having structural problems and institutional risks of adapting regulations—maintains the effects of weaknesses and inadequacies for years that provide insight on the limits to company growth and sustainable investment. This situation has been more severely disrupted by the crisis.

By all means SMEs, being small in size, cannot bear the entire financial risk of projects, as stated before. Therefore, only large sized firms are on the platform to establish an energy market. If they come together to form joint ventures with domestic firms or either established foreign energy companies the result might change. Without foreign capital, with merely the support of international financial institutions, the market will not be able to prosper because this is primarily a risk-based capital system that Turkey is trying to master. One of the most important difficulties that local companies face in Turkey is capital inadequacy, and therefore, they are insufficient to act as market-makers. Since 2005, international partners have seen some potential of a stable economy and good indicators, free from country risk, to invest and they are more willing to invest in this promising sector. At the same time for secured loans, it is important to assess the end-user payment default risk, energy company non-performance risk and bank's loan repayment default risk. These risks are to be appropriately addressed in order to ensure successful projects. What's more, energy projects need stable economic environment with low interest rate and inflation.

The energy projects in Turkey may be smaller than those in the industrial sector. To develop a prosperous energy market which determines counterparty and liquidity risk, it should be noted that macro-economic, micro-economic and financial factors are now subject to additional influences like global liquidity concerns. Therefore, the future Turkish energy market is somehow complicated made up of macro-economic, micro-economic and financial risks coming through both domestic and global means. Unfortunately, market performance is bound to market size problem, high interest rates, loan security problem which are severe constraints that might block the progress of the market. Especially, to maintain a lower credit risk of financing energy projects, it is a must to align with international energy companies and to seek possible business partnerships, securing third party financing from international finance organizations beyond WB such as the IFC, EEAF and the REEF. Beyond, the management of financial burdens there is a need of administration and enforcement of all the related legislation and requirements in order to obtain a secure future of energy in Turkey.

References

[1]   Athale, S. and Chavan, M. (2008) ESCOs: The Need of the Hour for Energy Efficiency in India. *The Bulletin on Energy Efficiency*, **8**, 34-35.
      http://www.docstoc.com/docs/26030648/ESCOs_-The-need-of-the-hour-for-Energy-Efficiency-in-India

[2]   Hopper, N., Goldman, C. and Mcwilliams, J. (2005) Public and Institutional Markets for ESCO Services: Comparing Programs, Practices and Performance. Ernest Orlando Lawrence Berkeley National Laboratory, Berkeley.
      http://emp.lbl.gov/sites/all/files/REPORT%20lbnl%20-%2055002.pdf

[3]   Bertoldi, P., Berrutto, V., de Renzio, M., Adnot, J. and Vine, E. (2003) How Are EU ESCOs Behaving and How to Create a Real ESCO Market? *Proceedings of the European Council for Energy Efficient Economy*, Stockholm, 2-7 June 2003, 909-916.
      http://www.eceee.org/library/conference_proceedings/eceee_Summer_Studies/2003c/Panel_5/5041bertoldi/paper

[4]   Osborn, J., Goldman, C., Hopper, C. and Singer, T. (2002) Assessing US ESCO Industry: Results from the NAESCO Database Project ACEEE Summer Study on Energy Efficiency in Buildings. Ernest Orlando Lawrence Berkeley National Laboratory, Berkeley.
      http://emp.lbl.gov/publications/assessing-us-esco-industry-results-naesco-database-project

[5]   Goldman, C. (2003) Overview of US ESCO İndustry: Recent Trends and Historic Performance. *International Workshop on Energy Efficiency Services Industries*, Shanghai, 8 September 2003, 1-33.

[6]   Silvoni, R. (2007) ESCO and Third Party Financing. *IInd Seminar of Information Campaigns*, Plovdiv, 5 March 2007, 1-14.
      http://www.seea.government.bg/documents/twinning/information_campaign_B_5_mar_06/information_campaign_2b.pdf

[7]   Sghaier, M. (2007) The Energy Audit in Tunisia and İts İnteraction with the ESCO Development Strategy. *NEEP, Workshop*: *Discovering Business Opportunities in the Energy Services Industry*. Northeast Energy Efficiency Partnerships (NEEP), Pasadena,
      http://www.neep.org.sa/en/downloads/workshops/w02/w02_19_sghaier_paper_(en).pdfCherail)

[8]   Sarkar, A. and Jas, S. (2010) Financing Energy Efficiency in Developing Countries—Lessons Learned and Remaining Challenges. *Energy Policy*, **38**, 5560-5571. http://dx.doi.org/10.1016/j.enpol.2010.05.001

[9]   Bertoldi, P., Boza-Kiss, B. and Rezessy, S. (2007) Latest Development of Energy Service Companies across Europe—A European ESCO Update. European Commission Joint Research Centre, Institute Environment and Sustainability, Ispra.
      www.energy.eu/publications/LBNA22927ENC_002.pdf

[10] Akman, U., Okay, E. and Okay, N. (2013) Current Snapshot of the Turkish ESCO Market. *Energy Policy*, **60**, 106-115.
      http://dx.doi.org/10.1016/j.enpol.2013.04.080

[11] Okay, E., Okay, N. and Akman, U. (2012) Turkey Chapter. In: Langlois, P. and Hansen, S.J., Eds., *World ESCO Outlook*, The Fairmont Press, Lilburn, 396-403.

[12] EIE (Elektrik İşleri Etüd) (2013) General Directorate of Renewable Energy of Ministry of Energy and Natural Resources. http://www.eie.gov.tr

[13] Benli, H. (2013) Potential of Renewable Energy in Electrical Energy Production and Sustainable Energy Development of Turkey: Performance and Policies. *Renewable Energy*, **50**, 33-46. http://dx.doi.org/10.1016/j.renene.2012.06.051

[14] TSKB (Türkiye Sınai Kalkınma Bankası, Industrial Development Bank of Turkey) (2013) Energy Efficiency.
      http://en.tskbenerjiverimliligi.com/energy-effiency-and-tskb/sources-of-funding-and-criteria.aspx

[15] Ürge-Vorsatz, D., Langlois, P. and Rezessy, S. (2004) Why Hungary? Lessons Learned from the Success of the Hungarian ESCO İndustry. *Proceedings of the* 2004 *ACEEE Summer Study on Energy Efficiency in Buildings*, Asilomar, 2004, 348-353.
http://www.eceee.org/library/conference_proceedings/ACEEE_buildings/2004/Panel_6/p6_30/paper

[16] Bertoldi, P. and Rezessy, S. (2005) Energy Service Companies in European Countries: Status Report 2005. European Commission, Ispra.

[17] Kostka, G. and Shin, K. (2011) Energy Service Companies in China: The Role of Social Networks and Trust. Frankfurt School—Working Paper Series.
http://www.frankfurt-school.de/clicnetclm/fileDownload.do?goid=000000302332AB4

[18] Morel, A. (2008) Terminal Evaluation of Project on "Developing Financial Intermediation Mechanisms for Energy Efficiency Projects in Brazil, China and India". UNEP (United Nations Environment Programme).
http://www.unep.org/eou/Portals/52/Reports/Energy_Efficiency_in_Brazil-China-India.pdf

[19] Lamers, P., Vollrad, K. and Anja, K. (2008) International Experiences with the Development of ESCO Markets. CONAE & GTZ, Berlin.
http://www.gtz.de/de/dokumente/en-International-Experience-Developing-ESCO-Markets.pdf

[20] Bannai, M., Tomita, Y., Ishida, Y., Miyazaki, D., Akisawa, A. and Kashiwagi, T. (2007) Risk Hedging against the Fuel Price Fluctuation in Energy Service Business. *Energy*, **32**, 2051-2060.
http://dx.doi.org/10.1016%2fj.energy.2007.05.003

[21] Okay, E., Okay, N., Konukman, A.E.S. and Akman, U. (2008) Views on Turkey's Impending ESCO Market: Is It Promising? *Energy Policy*, **36**, 1821-1824. http://dx.doi.org/10.1016/j.enpol.2008.02.024

[22] IFC (2006) Finance For Energy Efficient Street Lighting.
www.ifc.org/ifcext/gfm.nsf/AttachmentsByTitle/FMS-EO-EEF-EESL/$FILE/FMS-EO-EEF-EESL.pdf

[23] EBRD (European Bank for Reconstruction and Development) (2005) Financing ESCOs in Transition Economies.
http://www.docstoc.com/docs/27595905/Financing-ESCOs-in-Transition-Economies-%5BEBRD---Energy-efficiency%5D

[24] Shen, B., Price, L., Wang, J. and Li, M. (2012) China's Approaches to Financing Sustainable Development: Policies, Practices, and Issues. Ernest Orlando Lawrence Berkeley National Laboratory, Berkeley.
http://china.lbl.gov/sites/all/files/lbl-5579e-green-finance-wiresjune-2012.pdf

[25] BDA (Bangalore Development Authority) (2004) Municipal Street Lighting Energy Saving Project.
http://www.teriin.org/nss/PDDs/project4.pdf

[26] Vine, E. (2005) An International Survey of the Energy Service Company (ESCO) Industry. *Energy Policy*, **33**, 691-704.
http://dx.doi.org/10.1016/j.enpol.2003.09.014

[27] Bertoldi, P., Rezessy, S. and Vine, E. (2006) Energy Service Companies in European Countries: Current Status and a Strategy to Foster Their Development. *Energy Policy*, **34**, 1818-1832. http://dx.doi.org/10.1016/j.enpol.2005.01.010

[28] IFC (International Finance Corporation) (2013) Hungary Energy Efficiency: An Innovative GEF Project. FTF Brazil.
http://3countryee.org/RoundTable/HungaryEE.pdf

[29] BIS (Bankingfor International Settlements) (2013) Progress report on Basel III implementation.
http://www.bis.org/publ/bcbs232.pdf

[30] Narbel, P.A. (2013) The Likely Impact of Basel III on a Bank's Appetite for Renewable Energy Financing. NHH (Norges Handelshøyskole), Bergen.
http://www.nhh.no/Files/Filer/institutter/for/dp/2013/1013.pdf

[31] OECD (2013) Country Risk Classification. http://www.oecd.org/tad/xcred/crc.htm

[32] UniCredit (2013) Credit Risk. https://www.unicreditgroup.eu/en/investors/risk-management/credit.html

[33] Pykhtin, M. and Steven, Z. (2007) A Guide to Modelling Counterparty Credit Risk. *GARP Risk Review*, **37**, 16.
http://mhderivativesolutions.com/wp-content/uploads/2013/09/ssrn-id1032522.pdf

[34] Kleindorfer, P. (2010) Risk Management for EE Projects in Developing Countries. INSEAD Working Paper.
http://www.insead.edu/facultyresearch/research/doc.cfm?did=43993

[35] IET (Institute for Energy and Transport of European Commision) (2013) Energy Performance Contracting.
http://iet.jrc.ec.europa.eu/energyefficiency/european-energy-service-companies/energy-performance-contracting

[36] Thomas, J. (2009) Key Risk and Structuring Provisions for Bankable Transactions, Energy Project Financing: Resources and Strategies for Success. The Fairmont Press, Lilburn.

[37] Woodroof, A.E. (2009) Financing Energy Management Projects, Financing Energy Management Projects: Resources and Strategies for Success. The Fairmont Press, Lilburn.

[38] de Boer, S. (2011) Views on the Emerging Dutch ESCO Market: Can It Become Successful? Utrecht University,

Utrecht. http://www.eukn.org/dsresource?objectid=326773&type=org

[39] Central Bank of the Republic of Turkey, 2013. www.tcmb.gov.tr

[40] Özatay, F. (2006) Cari İşlemler Dengesine İlişkin İki Yapısal Sorun ve Mikro Reform Gereği. *Uluslararası Ekonomi ve Dış Ticaret Politikaları*, **No. 1**, 38-50.

[41] Akçay, C. and Üçer, M. (2008) A Narrative on the Turkish Current Account. *The Journal of International Trade and Diplomacy*, **2**, 211-238.

[42] Yükseler, Z. (2011) Türkiye'nin Karşılaştırmalı Cari İşlemler Dengesi ve Rekabet Gücü Performansı (1997-2010 Dönemi). TCMB Booklets, Central Bank Of Turkey, Ankara.

[43] Okay, E., Baytar, R. and Sarıdoğan, E. (2012) The Effects of Exchange Rate Changes on the Main Macroeconomic Variables in the Turkish Economy: An Econometric Analysis. *İktisat İşletme ve Finans Dergisi*, 27, 79-101.

[44] Soytas, U. and Sari, R. (2009) Energy Consumption, Economic Growth, and Carbon Emissions: Challenges Faced by an EU Candidate Member. *Ecological Economics*, **68**, 1667-1675. http://dx.doi.org/10.1016/j.ecolecon.2007.06.014

[45] KOSGEB (2013) Energy Efficiency Project in the SMES in Turkey. http://www.kosgeb.gov.tr/Pages/UI/Baskanligimiz.aspx?ref=107

[46] Turkish Commercial Code No. 6102, 2013. www.resmigazete.gov.tr/eskiler/2011/02/20110214-1-1.htm

[47] Okay, N. and Akman, U. (2010) Analysis of ESCO Activities Using Country Indicators. *Renewable and Sustainable Energy Reviews*, **14**, 2760-2771. http://dx.doi.org/10.1016/j.rser.2010.07.013

7

Effects of Energy Production and Consumption on Air Pollution and Global Warming

Nnenesi Kgabi[1], Charles Grant[2], Johann Antoine[2]

[1]Department of Civil and Environmental Engineering, Polytechnic of Namibia, Windhoek, Namibia
[2]International Centre for Environmental and Nuclear Sciences, University of the West Indies, Kingston, Jamaica
Email: nkgabi@polytechnic.edu.na

Abstract

In this study, different fuel combinations that can be adopted to reduce the level of air pollution and GHG emissions associated with the energy generation are assessed; and the air pollution and global warming effects of the Jamaican electricity generation fuel mix are determined. Based on the energy production and consumption patterns, and global warming potentials, the authors conclude that: an increase in energy consumption and production yields an increase in GHGs and other major pollutants; choice of the fuel mix determines the success of GHG emissions reductions; and there is no single fuel that is not associated with GHG or other air pollution or environmental degradation implications.

Keywords

Energy Generation, Energy Utilization, Air Quality Implications, GHG Emissions, Jamaica

1. Introduction

The harvesting, processing, distribution, and use of fuels and other sources of energy have major environmental implications including land-use changes due to fuel cycles such as coal, biomass, and hydropower, which affect both the natural and human environment. Energy systems carry a risk of routine and accidental release of pollutants [1]. Greenhouse gas (GHG) and air pollutant emissions share the same sources—transport, industry, commercial and residential areas [2]. All these sources depend on production, distribution and utilization of energy for their daily activities.

The gases included in GHG inventories are the direct GHGs: namely, carbon dioxide (CO_2), methane (CH_4), nitrous oxide (N_2O), hydrofluorocarbons (HFCs), perfluorocarbons (PFCs) and sulphur hexafluoride (SF6), and the indirect GHGs: non-methane volatile organic compounds (NMVOC), carbon monoxide (CO), nitrogen oxide (NOx), and sulphur dioxide (SO_2) [2].

Jamaica has no known primary petroleum or coal reserves and imports all of its petroleum and coal requirements. Domestic energy needs are met by burning petroleum products and coal and renewable fuel biomass (*i.e.*, biogases, fuel wood, and charcoal) and using other renewable resources (e.g., solar, wind and hydro). In 2008, approximately 86 percent of the energy mix was imported petroleum, with the remainder coming from renewables and coal [3]. Electricity is generated primarily by oil-fired steam, engine driven, and gas turbine units. Smaller amounts of electricity are generated by hydroelectric and wind power. Use of solar energy is negligible, and the option for nuclear energy has not been exploited.

The increase in GHG emissions, with the country just a few years from the global warming tipping point is evident. Carbon dioxide emissions increased from 9531 Gg in 2000 to 13,956 Gg in 2005, methane emissions from 31.1 Gg in 2000 to 41.9 Gg in 2005. Nitrous oxide emissions also increased although in smaller quantities. Emissions from the electricity generation source category between 2000 and 2005 ranged from 2977 Gg to 3365 Gg for CO_2, 0.116 Gg to 0.132 Gg for methane, and 0.023 Gg to 0.026 Gg for N_2O [4].

The objective of this study was to assess different fuel combinations that can be adopted to reduce the level of air pollution and GHG emissions associated with the energy. The study bears significance to almost all countries that, due to the pressure of high energy demand, tend to settle for any available energy source without considering the environmental effects.

2. Methods

Desktop study methods were used to source data for this secondary research. Information relating to energy generation and utilization in Jamaica was accessed outside organizational boundaries, mainly from online sources, research journals, professional bodies/organisations and government published data and reports. The data acquisition approach was purposive and mostly "cherry-picking", *i.e.*, based on keyword searches, footnote chases, citation searches or forward chains, journal runs, and to some extent author searches.

Content analysis methods for drawing conclusions included noting patterns, themes and trends, making comparisons, building logical chain of evidence and making conceptual/theoretical coherence. The content analysis yielded some descriptive data giving a detailed picture of the energy generation, air pollution and climate change in Jamaica.

3. Results and Discussion

Electricity generation, transmission, and distribution are associated with GHG emissions like carbon dioxide (CO_2), and smaller amounts of methane (CH_4) and nitrous oxide (N_2O). These gases are released during the combustion of fossil fuels, such as coal, oil, and natural gas, to produce electricity. Less than 1% of greenhouse gas emissions from the electricity sector come from sulfur hexafluoride (SF_6), an insulating chemical used in electricity transmission and distribution equipment [5].

3.1. Electricity Consumption

Consumption of electricity has direct GHG emission implications for the company/organization generating the electricity, and indirect implications for the consumer. **Table 1** shows the annual increase in GHG emissions with increase in electricity consumption. The main electric utility related gases are: GHGs—CO_2, CH_4, N_2O, SF_6; and Air Pollutants—CO, SO_2, NOx, NMVOCs.

Electricity sales data was obtained from World Bank, Benchmarking data of the electricity distribution sector in Latin America and the Caribbean Region 1995-2005; and the emission factors: CO_2—0.819 tons CO_2/MWh; CH_4—0.03716 kg CO_2/MWh; N_2O—0.00743 ton CO_2/MWh were used for calculation of the GHGs.

3.2. Electricity Generation

The annual fuel use for electricity generation between 2000 and 2005 ranged from 5,159,687 to 4,811,726 million barrels of heavy fuel oil and from 725,158 to 1,794,870 million barrels of diesel oil. Global warming potential of the fuel use are summarized in **Table 2**.

Table 1. Annual electricity sales and GHG emissions (data source for electricity sold [6]; source for emission factors used in the calculations [7]).

	Electricity Sold (MW h)	Methane (tonCO$_2$e)	Nitrous Oxide (tonCO$_2$e)	Carbon Dioxide (tonCO$_2$e)	Total GHG Emissions
2001	2,793,375	103.8	20754.8	2,287,774	2,308,633
2002	2,896,547	107.6	21521.3	2,372,271	2,393,900
2003	2,998,345	111.4	22277.7	2,455,644	2,478,033
2004	2,975,509	110.6	22,108	2,436,942	2,459,161
2005	3,055,154	113.5	22699.8	2,502,171	2,524,984
Average	2,943,786	109.38	21872.32	2410960.4	2,432,942
SD	101,539	3.769	754.431	83160.688	83918.9

Table 2. GHG emissions from combustion of fuel during electricity generation (data source for electricity sold [6]; source for emission factors used in the calculations [7]).

		Consumption (million barrels)	CO$_2$ ($\times 10^6$ tons CO$_2$e)	CH$_4$ ($\times 10^6$ ton CO$_2$e)	N$_2$O ($\times 10^6$ tons CO$_2$e)	Total GHG ($\times 10^6$ tons CO$_2$e)
2000	Fuel Oil	5,159,687	2513	2.16	6.96	2522
	Diesel Oil	725,158	306,322	104	2202	308,627
2005	Fuel Oil	4,811,726	874	0.75	2.42	877
	Diesel Oil	1,794,870	758,190	257	5450	763,897

The emission factors used above were obtained from DEFRA [7] as follows: Diesel = 2.6569, 0.0009, 0.0191, 2.6769 kg CO$_2$e/L; and Fuel Oil = 0.26729, 0.00023, 0.000074, 0.26826 kg CO$_2$ e/kWh; for CO$_2$, CH$_4$, N$_2$O, and CO$_2$e respectively. Other conversions used include 1 barrel = 158.99 liters; and 1 kWh = 0.00009 tonne oil equivalent.

Choice of the right fuel mix for electricity generation determines the amount of air pollutants and GHGs released into the atmosphere. The current electricity fuel mix of Jamaica is fuel oil (71%), diesel oil (24%) and 5% renewable. **Figure 1** shows the fuel mix used in 2007 based on the installed capacity by energy sources (MWh) data obtained from the United States Energy Information Administration (EIA) [8].

3.3. The Energy Fuel Mix

Jamaica's National Energy Policy 2009-2030: contribution of fuel mix to electricity generation mix is summarized in **Figure 2**.

The 3.3 percent increase in annual electricity generation (GWh) over the period (1998-2009), moving from 2950 GWh in 1998 to 4214 GWh in 2009; was used as baseline to estimate possible implications of the National Energy Policy on GHG emissions as shown in **Figure 3**.

Contribution by each fuel type is shown in **Figure 4**, emphasizing the importance of a correct fuel mix in reduction of GHG emissions.

Conversions used include: Natural Gas—CO$_2$ = 0.18483, CH$_4$ = 0.00027, N$_2$O = 0.00011, CO$_2$e = 0.18521 kg COe/unit; Coal-CO$_2$ = 0.32360, CH$_4$ = 0.00006, N$_2$O = 0.00282, CO$_2$e = 0.32648 kg COe/unit; and LPG-CO$_2$ = 0.21419, CH$_4$ = 0.00010, N$_2$O = 0.00025, CO$_2$e = 0.21455 kg COe/unit.

3.4. Electricity Distribution

Figure 5 shows annual percentages of losses that occur during distribution of electricity in Jamaica. The data used was obtained from the United States Energy Information Administration (EIA). Electrical transmissions and distribution systems contribute significantly to emissions of sulfur hexafluoride (SF$_6$), which is also a GHG. The losses during distribution also add to the emissions.

3.5. Air Quality Implications

Trace gases and aerosols impact climate through their effect on the radiative balance of the earth. Trace gases such as greenhouse gases absorb and emit infrared radiation which raises the temperature of the earth's surface causing the enhanced greenhouse effect. Aerosol particles have a direct effect by scattering and absorbing solar radiation and an indirect effect by acting as cloud condensation nuclei. Atmospheric aerosol particles range from dust and smoke to mists, smog and haze [10]. **Figure 6** gives the average air pollutant contribution by coal and petroleum products.

In addition to emission of GHGs, fuel combustion affects air quality. Combustion of 1 kg of coal results in emission of 19 g SO_2, 1.5 g NOx, 5 g VOCs, 4.1 g PM_{10}, 14.7 g TSP, 187.4 g CO and 0.0134 g benzene; while 1 kg of petroleum products emits 0.01 g SO_2, 1.4 g NOx, 0.5 g VOCs, 0.07 g PM_{10}, 0.07 g TSP, and 13.6 g CO into the atmosphere [11].

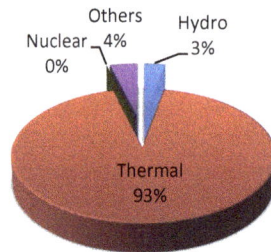

Figure 1. 2007 electricity generation fuel mix (data source: [8]).

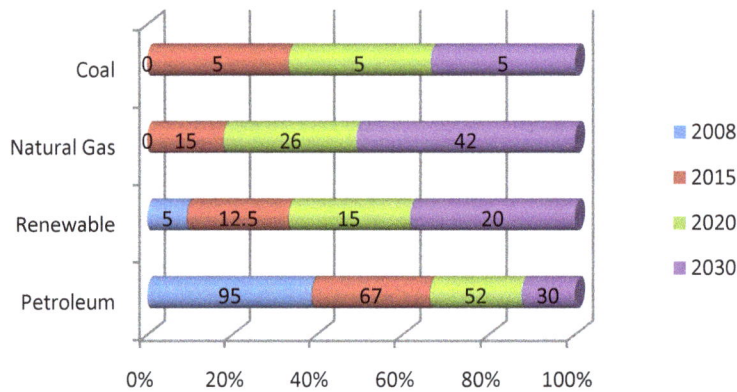

Figure 2. Electricity generation fuel mix proposed in the National Energy Policy (data source: [9]).

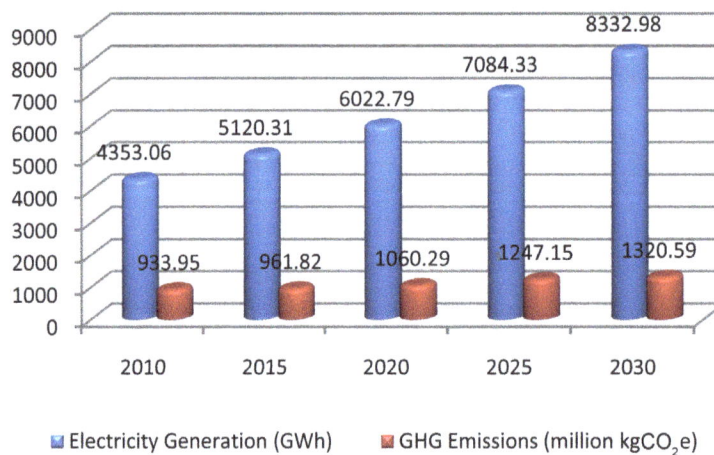

Figure 3. Estimate energy production and the potential GHG emissions (data source: [9]).

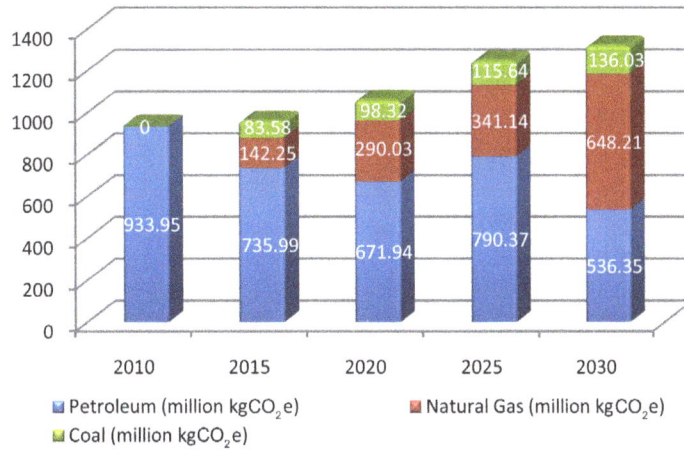

Figure 4. Contribution of the proposed fuels to GHG emissions (data source: [2], [7]).

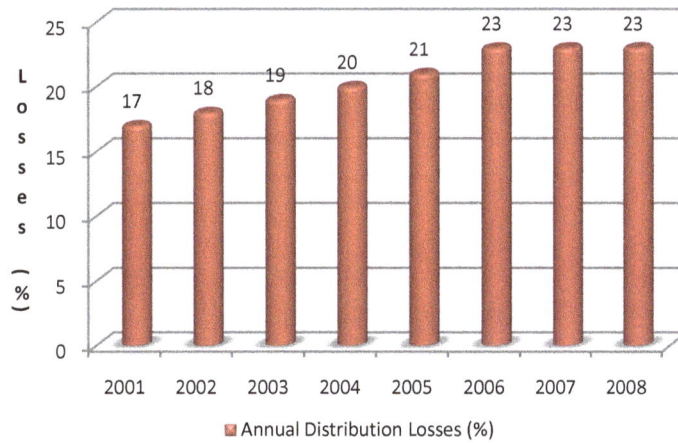

Figure 5. Electricity distribution losses (data source: [8]).

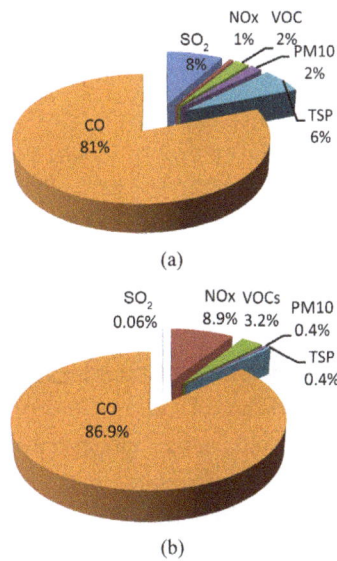

(a)

(b)

Figure 6. (a) Coal air pollutants; (b) Petroleum air pollutants (data source: [11]).

On combustion, fuel oil also produces primarily carbon dioxide and water vapour, but also smaller quantities of particulate matter and oxides of nitrogen and sulphur (OUR, 2012).

The potential environmental problem associated with coal is the formation of acidic effluents due to pyrite oxidation and the consequent mobilization of environmentally hazardous trace elements, which are mainly associated with the sulphide group of minerals in coal [12]. With combustion, these amounts of the SOx gases are emitted into the atmosphere.

4. Conclusions

The different fuel combinations (including coal, petroleum products, and natural gas) that can be adopted to reduce the level of air pollution and GHG emissions associated with the energy were assessed. This study has shown that: 1) choice of the fuel mix determines the success of GHG emissions reductions; and 2) there is no single fuel that is not associated with GHG or other air pollution or environmental degradation implications. The usual increase in GHG emissions with increase in energy consumption and production was observed.

Given the increasing energy demand and the environmental implications of the fuel mix options discussed, it may be necessary to also explore the nuclear energy option. Nuclear power plants do not require a lot of space when compared to equivalent wind or solar farms. The nuclear energy does not contribute to carbon emissions (no CO_2 is given out) thus does not cause global warming. Production and consumption of the nuclear energy do not produce smoke particles to pollute the atmosphere. The great advantage of nuclear power is its enormous energy density, several million times that of chemical fuels. Even without recycling, one kilogram of oil produces about 4 kWh; a kg of uranium fuel generates 400,000 kWh of electricity. This also reduces transport costs (although the fuel is radioactive and therefore each transport that does occur is expensive because of security implications). Furthermore, it produces a small volume of waste.

Acknowledgements

Support by the Academy of Sciences for the Developing World (TWAS), Polytechnic of Namibia, International Centre for Environmental and Nuclear Sciences (ICENS) and Faculty of Science and Technology, University of the West Indies (Mona Campus) is highly acknowledged.

References

[1]   Holdren, J.P. and Smith, K.R. (2000) Energy, the Environment, and Health. In: *World Energy Assessment: Energy and the Challenge of Sustainability*, 63-110.

[2]   IPCC (2006) Guidelines for National Greenhouse Gas Inventories. http://www.ipcc-nggip.iges.or.jp/public/2006gl/index.html

[3]   Office of Utilities Regulation (OUR) (2012) Sulphur Content of Fuel Oil Used for Power Generation. www.cwjamaica.com/-office.our

[4]   Jamaica Productivity Centre (2010) Generation and Distribution of Electricity in Jamaica: A Regional Comparison of Performance Indicators.

[5]   United States Environmental Protection Agency (USEPA) (2012) Sources of Greenhouse Gas Emissions. http://www.epa.gov/climatechange/ghgemissions/sources/electricity.html

[6]   World Bank (2013) Benchmarking Data of the Electricity Distribution Sector in Latin America & the Caribbean Region 1995-2005. http://info.worldbank.org/etools/lacelectricity/home.htm

[7]   DEFRA (2012) 2012 Guidelines to DEFRA/DECC's GHG Conversion Factors for Company Reporting. AEA for the Department of Energy and Climate Change (DECC) and the Department for Environment, Food and Rural Affairs (DEFRA).

[8]   United States Energy Information Administration (EIA) (2012). http://www.iea.org/stats/index.asp

[9]   Ministry of Energy and Mining (MEM) (2009) Jamaica's National Energy Policy 2009-2030.

[10]   IPCC (2001) Climate Change 2001: The Scientific Basis, Contribution of Working Group I to the Third Assessment Report of the Intergovernmental Panel on Climate Change.

[11]   Friedl, A., *et al.* (2004) Air Pollution in Dense Low-Income Settlements in South Africa. Royal Danish Embassy, Department of Environmental Affairs and Tourism, 2008.

[12]   Garcia, A.B. and Martinez-Tarazona, M.R. (1993) Removal of Trace Elements from Spanish Coals by Flotation. *Fuel*, **72**, 329-335. http://dx.doi.org/10.1016/0016-2361(93)90050-C

8

A Stable Energy Saving Adaptive Control Scheme for Building Heating and Cooling Systems

Sumera I. Chaudhry, Manohar Das

Department of Electrical and Computer Engineering, Oakland University, Rochester, USA
Email: sichaudh@oakland.edu, das@oakland.edu

Abstract

This paper presents a stable, nonlinear, adaptive control scheme for building heating and cooling systems. The proposed controller utilizes the principle of adaptive one step ahead control and aims at reducing the energy consumed for heating or cooling a building. The design steps are discussed in details and a proof of global stability is also provided. Also, the performance of the proposed controller is demonstrated on a simulated building thermal model.

Keywords

Building HVAC Control, Adaptive Control, Energy Saving Control

1. Introduction

According to EPA estimates, the cost of heating, ventilation and cooling (HVAC) of residential and commercial buildings constitute about 50% of the total electrical energy consumed in USA. Thus, minimization of this HVAC energy consumption can be very beneficial not only because of the resulting cost savings but also because of the reduction of carbon footprints of buildings. In recent years, the installation of smart meters in buildings and advent of smart grid technology have provided an impetus to development of advanced control schemes for reducing the energy consumption and maintaining the thermal comfort inside a building. A variety of such advanced control schemes have appeared in the literature [1]-[7]. Some of these claim to be adaptive, efficient, and helpful in reducing the energy consumption while maintaining the occupants' comfort level. Examples include classical PID controller and PID plus fuzzy logic controller [1] [2], optimal, adaptive and intelligent controllers [2], adaptive neuro-fuzzy inference system (ANFIS) and artificial neural network (ANN) based controllers, and fuzzy logic (FL) controllers [3]. Also, other approaches include model predictive control (MPC) [4]

[5], and predicted mean vote (PMV) based adaptive or PID controller [6]-[8]. Till this date, however, very little efforts have been made in employing truly adaptive techniques that learn the nonlinear thermal characteristics of a building and its environment, and presenting a nonlinear adaptive controller backed up by a proof of globally stability. This paper is aimed at addressing these concerns.

Notice that adaptive controllers are ideally suited for controlling the heating/cooling system of a building, because such controllers can estimate the building's physical parameters and occupancy levels online and adjust the control signals accordingly. The main difficulty in designing such controllers stems from the fact that a building thermal system is usually characterized by a bilinear model. In view of this, we investigate the application of adaptive one step ahead (OSA) and weighted one step ahead (WOSA) control schemes [9] here, because such controllers are easy to design and capable of controlling both linear and bilinear systems quite well. The contributions of this paper are threefold. First we investigate in details the application of adaptive one step a head (OSA) and weighted one step ahead (WOSA) control schemes for controlling the thermal system of a building. Then we investigate the criteria for global stability of the closed loop system, and present a detailed proof of the same. Finally, we present results of a simulation study that compares the performance of the adaptive OSA and WOSA controllers with that of a simple fuzzy control scheme, and show that OSA and WOSA controllers are capable of reducing the heating/cooling energy consumption in a building.

The organization of this paper is as follows. Section 2 presents a brief overview of the objectives and methodology. A dynamical model of a building's thermal system is described briefly in Section 3. A description of OSA and WOSA controllers and a discussion of parameter estimation issues are presented in Section 4. A detailed proof of global stability of the closed loop system is presented in Section 5. Then Section 6 presents results of some simulation studies, and finally, some concluding remarks are given in Section 7.

2. An Overview of Objectives and Methodology

The aim of this paper is to present a scheme for controlling the indoor temperature of a building in a desired way using a globally stable control scheme that also exhibits good tracking. Such a controller is based on a simple nonlinear dynamical model of the thermal system of a building. The parameters can be different for different buildings and can be time varying, and in most cases we have no knowledge of the parameter values. Therefore, due to the nonlinear nature of the dynamical model and the unknown parameters, an adaptive control approach is used to fulfill the objective.

The goal of a smart building temperature control system is to maintain the indoor temperature of a building by achieving an optimum tradeoff between heating/cooling energy consumption and occupants' comfort. The overall heating/cooling energy consumed in a building can be divided into two parts: 1) energy consumed to maintain the current conditions, and 2) energy consumed to raise (or lower) temperature to a different level. The latter becomes an important part of the overall energy saving strategy because during favorable outdoor conditions, periods of low occupancy levels and at nights, the target temperature can be lowered (during winter season) or raised (during summer season) to save energy, and subsequently brought back to the desired level whenever desired. However, in doing so, the reference temperature profile needs to be chosen carefully by avoiding steep gradients (to reduce energy consumption) and at the same time maintaining a desired comfort level. The proposed adaptive control scheme involves measurement of inputs and outputs of the system, estimation of unknown system parameters using a recursive least squares (RLS) parameter estimation algorithm, and computation of a control signal based on the estimated parameter values. The design of both OSA and WOSA controllers is discussed and their performance is studied.

3. Dynamical Model of a Building's Thermal System

At the outset, it should be noted that since the heating and cooling dynamics are very similar, we discuss only the heating dynamics here for the sake of brevity. Following the footsteps of Calvino et al. [7] and IBPT toolbox [10], a simplified dynamical model of a building's thermal system during a heating season can be described by the following equation:

$$\frac{c_i \mathrm{d}\theta_a(t)}{\mathrm{d}t} = \frac{c_f \Delta\theta_f}{\Delta\theta_n}\dot{m}(t)\big(\theta_{\mathrm{fav}} - \theta_a(t)\big) + H_T\big(\theta_{\mathrm{out}}(t) - \theta_a(t)\big) + H_{\mathrm{gains}} \tag{1}$$

where $\theta_a(t)$ denotes the indoor air temperature at time t, $\theta_{out}(t)$ is the outdoor air temperature at time t, c_i denotes the total heat capacity of the indoor air mass and other objects inside the building, c_f is the specific heat of the warming carrier, $\Delta\theta_f$ denotes the temperature difference between the inlet and outlet ends of the heat exchanger, and $\Delta\theta_n$ denotes the difference between the average temperature of the heating medium and the indoor air temperature. Also, H_T denotes the global heat transfer coefficient of the building envelope and H_{gains} denotes the heat gains from various internal and external sources. A more detailed description of the above terms can be found in references [7] and [10].

Before presenting our control strategies, it would be convenient to make the following changes of variables for the sake of notational simplicity:

$$y = \theta_a, \ u = \dot{m}, \ y_{out} = \theta_{out}, \ y_{fav} = \theta_{fav}$$

Thus, Equation (1), which represents a simplified dynamical model of the building, now takes the following form:

$$\frac{dy}{dt}(t) = Au(t)\left(y_{fav} - y(t)\right) + B\left(y_{out}(t) - y(t)\right) + C \quad (2)$$

where $y(t)$ denotes the indoor temperature, $u(t)$ is the flow rate of the warming carrier and A, B and C denote the model parameters that are given by:

$$A = \frac{c_f \Delta\theta_f}{c_i \Delta\theta_n} \quad (3a)$$

$$B = \frac{H_T}{c_i} \quad (3b)$$

$$C = \frac{H_{gains}}{c_i} \quad (3c)$$

As considered in [1], we assume that the thermal losses due to ventilation are insignificant, but the convective part of all heat sources, such as the solar heat gains and the heat gains from the heating system or casual gains, are considered to be parts of the model equation. The following discrete time equation is derived from the system Equation (1) using a first order Euler approximation for $\frac{dy}{dt}$ with a sampling period T_s.

$$\frac{y(k+1) - y(k)}{T_s} = Au(k)\left(y_{fav} - y(k)\right) + B\left(y_{out}(k) - y(k)\right) + C \quad (4)$$

where k denotes the discrete time index (i.e., $k = 1, 2, 3, \cdots$) and the time instance, kT_s, is simply denoted by k. The above thermal model is characterized by three unknown parameters, namely, A, B and C.

4. Control of Indoor Temperature of a Building

To develop a control scheme for controlling the indoor temperature of a building governed by Equation (4), first thing to notice is that it represents a nonlinear system characterized by some unknown parameters. These parameters can vary from building to building, and in most cases we have no prior knowledge of the parameter values. In view of this, we propose to use a nonlinear, adaptive OSA and WOSA controllers [9]. Furthermore, it is important to establish that the resulting closed loop system is stable and exhibits good tracking behavior.

The proposed adaptive control scheme involves measurement of inputs and outputs of the system, estimation of unknown system parameters using a recursive least squares (RLS) parameter estimation algorithm, and computation of a control signal based on the estimated parameter values. First the design of fixed OSA and WOSA controllers is discussed and their performance is studied.

4.1. Fixed One Step Ahead Control Algorithm

The goal of a fixed OSA controller is to track a desired reference temperature profile, $y^*(k)$. Such a tracking is achieved by bringing the predicted output at time k + 1, *i.e.*, $y(k+1)$, to the desired value, $y^*(k+1)$ in one step. The feedback control law that achieves this goal is found by minimizing the following cost function based on a squared prediction error:

$$J_1(k+1) = \frac{1}{2}\left[y(k+1) - y^*(k+1)\right]^2 \tag{5}$$

The discrete time control law, obtained by differentiating $J_1(k+1)$ with respect to $u(k)$ and setting it to zero, is given by [9]:

$$\bar{u}(k) = \frac{y^*(k+1) - y(k) - T_s B\left[y_{\text{out}}(k) - y(k)\right] - T_s C}{\left[T_s A\left(y_{\text{fav}} - y(k)\right)\right]} \tag{6}$$

However, in view of the fact that any furnace is limited by its maximum heat delivery capacity, it is necessary to constrain the above control signal, $\bar{u}(k)$, and generate a constrained control signal, $u(k)$, as follows:

$$u(k) = \bar{u}(k) \quad \text{if} \quad 0 < \bar{u}(k) < u_{\text{max}} \tag{7a}$$

$$\text{else} \quad u(k) = u_{\text{max}} \quad \text{if} \quad \bar{u}(k) \geq u_{\text{max}} \tag{7b}$$

$$\text{else} \quad u(k) = 0 \quad \text{if} \quad \bar{u}(k) \leq 0 \tag{7c}$$

where u_{max} denotes the maximum low rate of the warming carrier.

4.2. Fixed Weighted One Step Ahead Control Algorithm

Since OSA controllers attempt to achieve zero tracking error in one step, they often require large control efforts, $u(k)$, which usually increases the overall energy consumption. To alleviate this drawback, a weighted OSA (WOSA) controller [9] is often found to be a good alternative. A WOSA controller attempts to achieve a trade-off between tracking error and control efforts. It minimizes the following cost function:

$$J_2(k+1) = \frac{1}{2}\left[y(k+1) - y^*(k+1)\right]^2 + \frac{\lambda}{2}\left[\bar{u}(k)\right]^2 \tag{8}$$

where the parameter, λ, controls the trade-off between tracking error and control efforts. A larger λ reduces control efforts at the cost of higher tracking error and vice versa.

The control law that minimizes $J_2(k+1)$ is given by [9]:

$$\bar{u}(k) = \frac{\left[T_s A\left(y_{\text{fav}} - y(k)\right)\right] * \left\{y^*(k+1) - y(k) - T_s B\left[y_{\text{out}}(k) - y(k)\right] - T_s C\right\}}{\left[T_s A\left(y_{\text{fav}} - y(k)\right)\right]^2 + \lambda} \tag{9}$$

which is once again constrained by u_{max} to generate a constrained control signal, $u(k)$, defined by Equations (7a)-(7c).

4.3. Adaptive OSA and WOSA Controllers

In an adaptive controller, the sampled measurements, $u(k)$ and $y(k)$, are used to estimate the model parameters, A, B and C in Equation (2), using a recursive parameter estimation method, such as recursive least squares (RLS). The estimated values of theses parameters are then used to compute the OSA/WOSA control signals.

4.3.1. Parameter Estimation
First we write model Equation (2) in the following form:

$$y(k+1) = \varphi(k)^{\text{T}} \theta^* \tag{10}$$

where

$$\varphi(k) = \left[T_s * u(k-1)\{y_{\text{fav}} - y(k-1)\} T_s * \{y_{\text{out}}(k-1) - y(k-1)\} T_s\right]^{\text{T}} \tag{11a}$$

$$\theta^* = \begin{bmatrix} A & B & C \end{bmatrix}^{\mathrm{T}} \tag{11b}$$

Next, the estimated value of θ^* is computed recursively using the following RLS algorithm:

$$\hat{\theta}(k) = \hat{\theta}(k-1) + \frac{P(k-2)\varphi(k-1)}{1 + \varphi(k-1)^{\mathrm{T}} P(k-2)\varphi(k-1)} \left[y(k) - \varphi(k-1)^{\mathrm{T}} \hat{\theta}(k-1) \right]; \quad k \geq 1 \tag{12a}$$

$$P(k-1) = P(k-2) - \frac{P(k-2)\varphi(k-1)\varphi(k-1)^{\mathrm{T}} P(k-2)}{1 + \varphi(k-1)^{\mathrm{T}} P(k-2)\varphi(k-1)} \tag{12b}$$

$$\hat{\theta}(0) = \begin{bmatrix} \gamma & 0 & 0 \end{bmatrix}^{\mathrm{T}} \tag{12c}$$

$$P(-1) = \sigma I \tag{12d}$$

where $\gamma > 0$ is a small number and $\sigma > 0$ is chosen to be large. Also, $\hat{A}(k)$ is always constrained to be non-negative by using a projection algorithm [9], i.e.,

$$\hat{A}(k) > \epsilon > 0 \quad \text{for all } k. \tag{12e}$$

Given an estimate $\hat{\theta}(k)$ of θ^*, we define the predicted output at time $k+1$ as:

$$\hat{y}(k+1) = \varphi(k)^{\mathrm{T}} \hat{\theta}(k) \tag{13}$$

4.3.2. Adaptive Control Algorithms

The adaptive OSA and WOSA controllers use the above estimate, $\hat{\theta}(k)$, to compute the control signal, $u(k)$, from the following adaptive versions of Equations (6) and (9):

For OSA:

$$u(k) = \frac{y^*(k+1) - y(k) - T_s \hat{B}(k)\left[y_{\mathrm{out}}(k) - y(k) \right] - T_s \hat{C}(k)}{\left[T_s \hat{A}(k)\left(y_{\mathrm{fav}} - y(k) \right) \right]} \tag{14}$$

For WOSA:

$$u(k) = \frac{\left[T_s \hat{A}(k)\left(y_{\mathrm{fav}} - y(k) \right) \right] * \left\{ y^*(k+1) - y(k) - T_s \hat{B}(k)\left[y_{\mathrm{out}}(k) - y(k) \right] - T_s \hat{C}(k) \right\}}{\left[T_s \hat{A}(k)\left(y_{\mathrm{fav}} - y(k) \right) \right]^2 + \lambda} \tag{15}$$

where $\hat{A}(k)$, $\hat{B}(k)$ and $\hat{C}(k)$ denote the estimated values of A, B and C, respectively, at time k.

5. Global Stability of the Closed Loop Adaptive Control System

A proof of global stability of the closed loop system governed by adaptive OSA controller is provided in this section under the following mild assumptions:

Assumption 1

The building thermal parameters, A, B and C have finite, positive values, i.e.,

$$0 < A < A_{\max} < \infty$$

$$0 < B < B_{\max} < \infty$$

$$0 < C < C_{\max} < \infty$$

Assumption 2

The room temperature, $y(k)$, is always less than the average temperature, y_{fav}, of the heat exchanger, i.e., $y(k) < y_{\mathrm{fav}}$ for all k.

The global stability of the closed loop system can now be established by first proving three Lemmas and finally proving Theorem 1 below.

Lemma 1

For the least squares algorithm describes by Equations (12a)-(12d), we have

$$\lim_{k \to \infty} \left| \hat{\theta}(k) - \hat{\theta}(k-m) \right| = 0, \ \text{for a finite } m \tag{16a}$$

$$\lim_{k \to \infty} \frac{e(k)}{\left[1 + a_1 \varphi(k-1)^{\mathrm{T}} \varphi(k-1)\right]^{1/2}} = 0 \tag{16b}$$

where a_1 is some constant and prediction error, $e(k)$, is given by

$$e(k) = y(k) - \hat{y}(k) = y(k) - \varphi^{\mathrm{T}}(k) \hat{\theta}(k-1) \tag{16c}$$

Proof
The proof of Lemma 1 can be found in Goodwin and Sin [9].
Next the boundedness of $y(k)$ is proved in Lemma 2 below.
Lemma 2
Consider the closed loop adaptive OSA control system governed by Equations (2), (14) and (7a)-(7c). This system is BIBO stable and therefore, $y(t)$ is bounded for all t.
Proof
Notice that at any $t \in \left[kT_s, (k+1)T_s\right]$, the closed loop system governed by Equation (2) can be rewritten as:

$$\frac{dy}{dt}(t) + \alpha(t)y(t) = Au(t)(y_{\text{fav}}) + B(y_{\text{out}}(t)) + C \tag{17}$$

where $u(t) = u(k)$, which is given by Equations (14) and (7a)-(7c), and $\alpha(t)$ is defined as

$$\alpha(t) = Au(t) + B \tag{18}$$

In view of (7a)-(7c) and assumptions 1 and 2, we get

$$0 \leq u(t) \leq u_{\max} \tag{19a}$$

$$0 < B < \alpha(t) < Au_{\max} + B \tag{19b}$$

First consider the homogeneous system associated with Equation (17), *i.e.*,

$$\frac{dy}{dt}(t) + \alpha(t)y(t) = 0 \tag{20}$$

The state transition matrix [11] of this homogeneous system is given by

$$\Phi(t, t_0) = e^{-\int_{t_0}^{t} \alpha(\tau) d\tau} \tag{21}$$

In view of (19b), we have

$$\left| \Phi(t, t_0) \right| = \left| e^{-\int_{t_0}^{t} \alpha(\tau) d\tau} \right| \leq \left| e^{-\int_{t_0}^{t} B d\tau} \right| = e^{-B(t - t_0)} \quad \text{for all } t \geq t_0 \tag{22}$$

In view of (22), the homogeneous system given by Equation (20) is uniformly asymptotically stable [11].
Next, consider the closed loop system given by Equation (17). Since its homogeneous system described by (20) is uniformly asymptotically stable and the excitation given by the right hand side of Equation (17) is bounded for all times (in view of Assumption 1 and Equation (19a)), the system governed by (17) is BIBO stable [14]. This implies $y(t)$ is bounded for all t. Next, the following Lemma establishes the boundedness of the prediction error, $e(k)$, defined by Equation (16c).
Lemma 3

$$\lim_{k \to \infty} y(k) - \hat{y}(k) = 0 \tag{23}$$

Proof
In view of Equation (19a) and Lemma (2), the regression vector, $\varphi(k)$ defined by Equation (11a) is bounded, *i.e.*,

$$\|\varphi(k)\| < M_1 < \infty \tag{24}$$

where M_1 is some constant. In view of Equation (16b) of Lemma 1 and (24), we get (23). Finally, a tracking result is provided in Theorem 1 below.

Tracking

In order to prove tracking properties of the adaptive OSA controller, we are going to assume (for the sake of simplicity) that $y^*(t)$ is a constant, y^*. Notice that this is a reasonable assumption because $y^*(t)$ is actually piecewise constant for all t.

Theorem 1

Subject to the assumptions 1 and 2, the proposed adaptive OSA scheme assures that

$$\lim_{k\to\infty} |y(k) - y^*| = 0 \tag{25}$$

Proof

We present an outline of the proof here, because it is very similar to the results presented in [12] [13] and details can be found there. First of all, notice that from Lemma 1, we get

$$\lim_{k\to\infty} |y(k) - \hat{y}(k)| = 0 \tag{26}$$

Thus, for a given η, there exists a k_0 such that for $k \ge k_0$,

$$\hat{y}(k) - \eta < y(k) < \hat{y}(k) + \eta \tag{27}$$

Define an interval,

$$L = [h_1, h_2] = [y^* - \eta, y^* + \eta] \tag{28}$$

From here on we assume $k \ge k_0$ and proceed to consider the following three possible cases for $u(k)$ given by (28) and (7a)-(7c):

(i) $0 < u(k) < u_{max}$ for $k = k_0 + j$, $j = 1, 2, 3, \cdots$

(ii) $u(k) = 0$ for $k = k_0 + j$, $j = 1, 2, 3, \cdots$

(iii) $u(k) = u_{max}$ for $k = k_0 + j$, $j = 1, 2, 3, \cdots$

First we show that there exists a $k_1 > k_0$ such that

$$y(k) \in L \quad \text{for } k > k_1 \tag{29}$$

Case (i) If (i) holds, then $y(k_0 + 1) = y^*$ and clearly $y(k_0 + 1) \in L$. Similarly, $y(k_0 + j) \in L$, $j = 1, 2, 3, \cdots$

Case (ii). If (ii) holds, then Equation (7c) implies $\bar{u}(k) \le 0$. This, in view of (14), yields $y(k_0 + j) \ge y^*$, $j = \{1, 2, 3, \cdots\}$. Thus, Equation (27) yields $y(k_0 + j) \ge y^* - \eta$. But $u(k) = 0$ means furnace is off, which implies $y(k_0 + j)$ decreases asymptotically to a steady state value determined by the outdoor temperature, $y_{out}(k)$. Thus, there exists a j' such that $y(k_0 + j') \le y^* + \eta$. Thus, $y(k_0 + j') \in L$. However, if the above sequence, $\{k_0 + j\}$ terminates at time k_1 before $y(k) \in L$, then furnace turns on, which returns us to case (i) and therefore, $y(k_1 + 1) \in L$.

Case (iii). If (iii) holds, then $y(k)$ increases due to heat output from the furnace. But $u(k) = u_{max}$ implies (from (7b)) $\bar{u}(k) \ge u_{max}$. This, in view of (14), yields $\hat{y}(k_0 + j) \le y^*$, $j = \{1, 2, 3, \cdots\}$. Thus, Equation (27) yields $y(k_0 + j) \ge y^* + \eta$, $j = \{1, 2, 3, \cdots\}$. In a similar fashion like in case (ii), we can once again prove that there exists a $k_1 > k_0$ such that $y(k) \in L$ for $k > k_1$.

From here on, by induction on (29), it follows that $y(k) \in L$ for some $k > k_1$. Finally, since η can be chosen arbitrarily, the above result also establishes (25).

Remarks

1) In the above analysis, noise is assumed to be absent. In presence of bounded noise, the analysis can be modified slightly and following arguments similar to [12] [13], it can be shown that there exists a k_1 such that for $k \ge k_1$,

$$\limsup_{k\to\infty} |y(k) - y^*| \le 2\Delta \tag{30}$$

where $|\text{noise}| < \Delta$ for all times.

2) For the adaptive OSA, the tracking error is zero, whereas for the adaptive WOSA, the tracking error is

proportional to the weight, λ, and inversely proportional to the square of the sampling time T_s, which can be shown easily by substituting control $u(k)$ in the equation for $y(k)$.

6. Simulation Results and Discussion

In this section, we present results of a simulation study to show the performance of OSA and WOSA controllers and also compare them with a simple fuzzy logic controller. These simulations are conducted on a simplified thermal dynamical model of a small family home that was studied by Calvino *et al.* [7]. However, in our case, this home is assumed to be located in Michigan, USA. The physical parameters of this home are assumed to be **Table 1**.

The simulations are carried out over a period of 24 hours, or equivalently 86,400 seconds. The outdoor temperature during this period is assumed to vary slowly in a sinusoidal fashion as follows:

$$\theta_{\text{out}}(t) = 5 + 0.75 \times \left(\sin\left(2\pi \times 0.00003 \times t \right) \right). \tag{31}$$

which represents the outdoor temperature variation on a typical winter day in Michigan, USA. However, the desired indoor temperature profile, $y^*(t)$, is assumed to be just a two-level signal consisting of a low level and a high level. It is set to be at the high level during early morning and evening hours, whereas it is lowered to the low-level setting during night time as well as office hours when the place is devoid of occupants.

The performance of an adaptive OSA controller is depicted in **Figure 1** and **Figure 2**. These simulations were carried out using two different values of maximum blower speed (u_{max}) of the furnace, 0.1 Kg/sec and 0.5 Kg/sec. **Figure 1(a)** and **Figure 1(b)** compare the actual indoor temperature with the desired one for the above settings of u_{max}. Although both show good tracking between the actual temperature and the desired one, we notice that the furnace with a higher u_{max} results in a faster rise of the actual temperature. **Figure 2(a)** and **Figure 2(b)** depict the behavior of the control signal, $u(t)$, in these two cases. As can be seen from these figures, the furnace with a higher u_{max} results in a smoother control signal. However, this improvement in performance has to be traded off against a higher capital cost for the high-capacity furnace.

The performance of an adaptive WOSA controller is depicted in **Figure 3** and **Figure 4**. Since we wish to show the effect of variation of the controller trade-off parameter, λ, we just use a lower capacity furnace with $u_{\text{max}} = 0.1$ Kg/s for this part of the study. However, the value of λ is chosen to be 0.1 and 0.05, respectively, for these two case. **Figure 3(a)** and **Figure 3(b)** compare the tracking errors for the above two cases. As can be seen from these figures, a smaller value of λ improves the tracking between the actual temperature and the desired one. **Figure 4(a)** and **Figure 4(b)** show the corresponding control signals. A comparison of these figures with **Figure 4(a)** and **Figure 4(b)** clearly shows that a WOSA controller substantially reduces the control efforts.

The performance of the above OSA and WOSA controllers is next compared with that of a simple fuzzy controller, described by the following equations:

$$u(t) = e(t) + \Delta e(t) \tag{32}$$

where the variables e (absolute error) and Δe (time variation of the error) is defined as

$$e(t) = y(t) - y^*(t) \tag{33}$$

$$\Delta e = e(t) - e(t-1) \tag{34}$$

Table 1. Physical parameters of the simulated building.

Total Internal Volume	$V = 271 \text{ m}^3$	Overall Heat Transmittance	$H_T = 183 \text{ W/K}$
Specific Heat of Air	$c_a = 1012 \text{ J/Kg} \cdot \text{K}$	Air Density	$\rho_a = 1.204 \text{ Kg/m}^3$
Temperature of Warming Carrier (Water)	$\theta_{f,\text{in}} = 75°\text{C}$	Specific Heat of Water	$c_f = 4186 \text{ J/Kg} \cdot \text{K}$
	$\theta_{f,\text{out}} = 65°\text{C}$	Thermal Gains	$Q_{\text{gains}} = 100 \text{ W}$

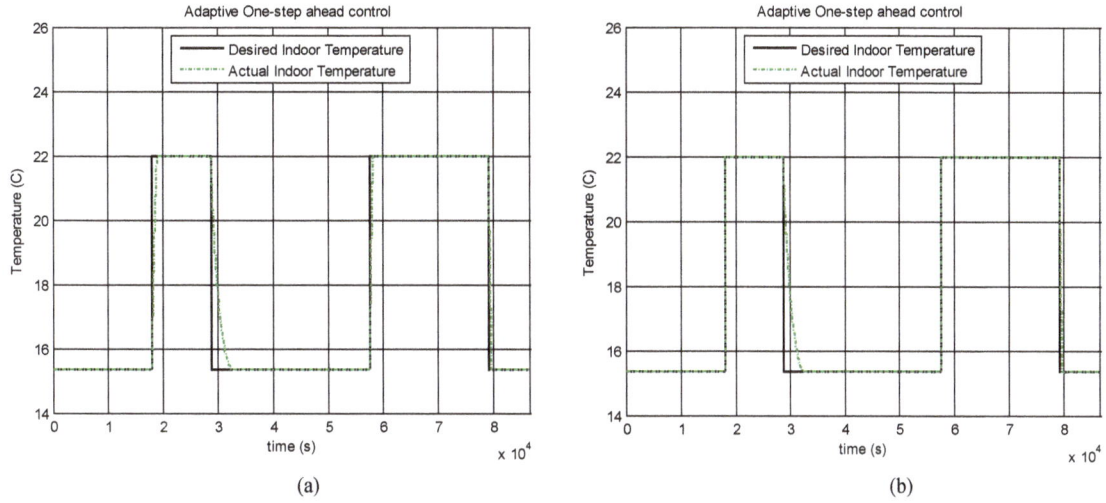

Figure 1. Comparison of actual and desired temperatures using adaptive OSA controller for (a) $u_{max} = 0.1$ Kg/s and (b) $u_{max} = 0.5$ Kg/s.

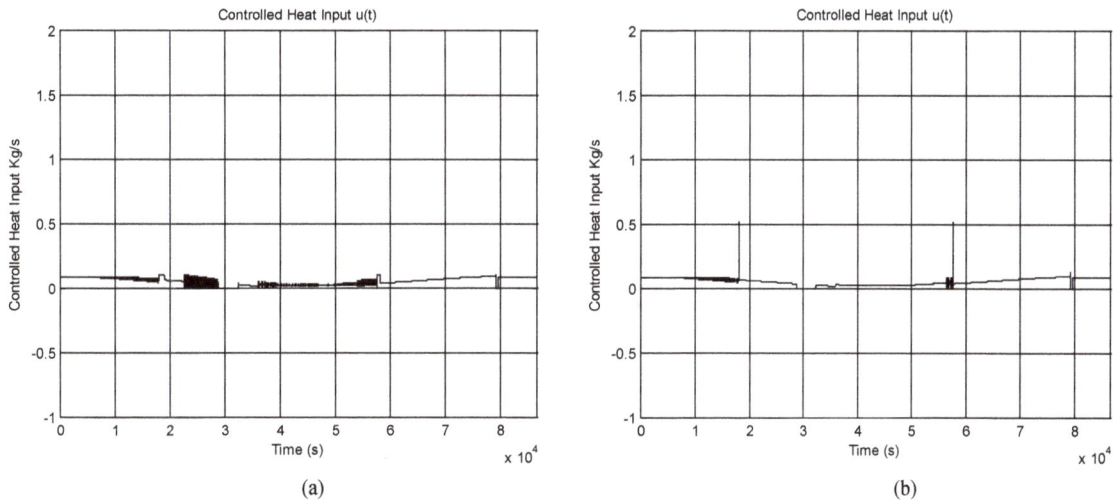

Figure 2. Behavior of adaptive OSA control signals for (a) $u_{max} = 0.1$ Kg/s and (b) $u_{max} = 0.5$ Kg/s.

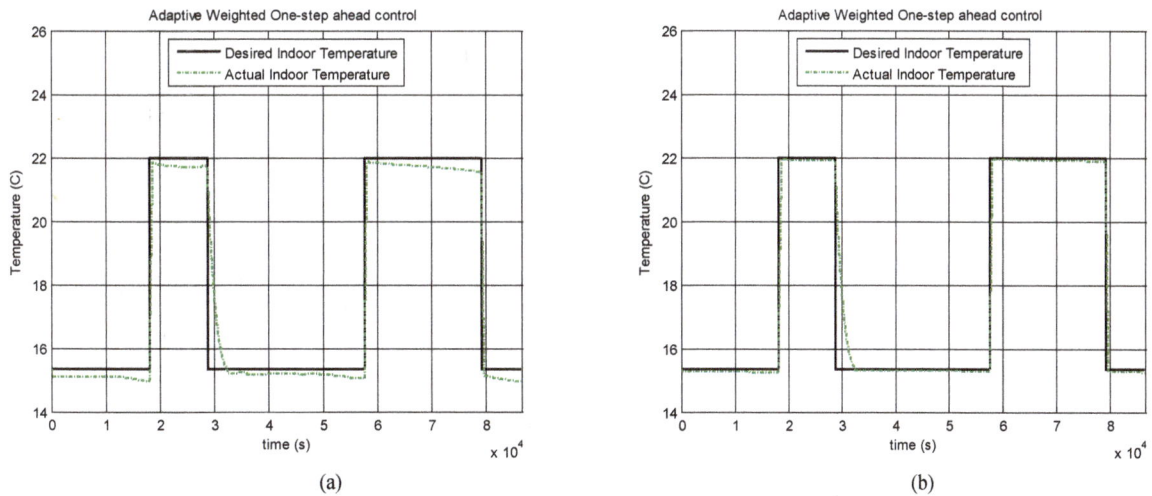

Figure 3. Comparison of actual and desired temperatures using adaptive WOSA controller for (a) $\lambda = 0.05$ and (b) $\lambda = 0.01$.

A fuzzy controller, consisting of 25 Sugeno-style decision rules, was designed using Matlab fuzzy logic toolbox and used in this part of our study. The performance of the above controller is depicted in **Figure 5** and **Figure 6**. **Figure 5(a)** shows the tracking performance for $u_{max} = 0.1$ Kg/seccc, which seems to be rather poor. The controller is unable to track the desired temperature profile because of the lower heat delivery capacity of the furnace. In view of this, u_{max} was increased to 0.5 Kg/sec to see if it would improve the tracking performance. As shown in **Figure 5(b)**, this indeed is the case, but the overall tracking performance is not as good as adaptive OSA and WOSA controllers. **Figure 6(a)** and **Figure 6(b)** depict the behavior of the control signals for the above two cases.

Finally, the energy consumptions of OSA, WOSA and fuzzy controllers using same reference temperature profile are compared in **Table 2**. It is evident from this table that both OSA and WOSA controllers can deliver significant energy savings per day, which becomes even more significant when the total number of cold days in a winter season is taken into consideration.

7. Conclusion

An adaptive one-step ahead control scheme and a weighted one-step ahead control scheme for a building's

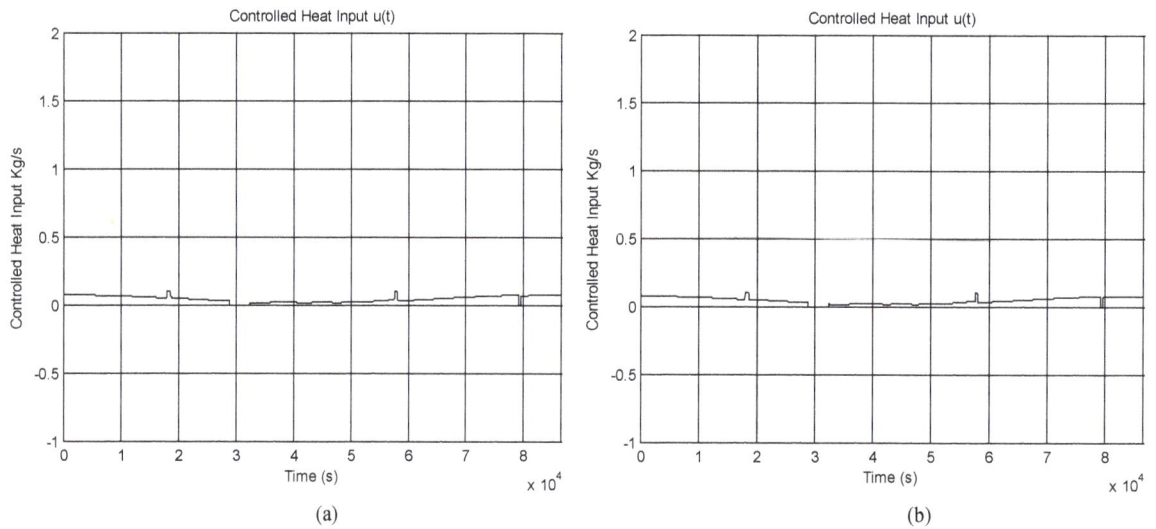

(a)   (b)

Figure 4. Behavior of adaptive WOSA control signals for (a) $\lambda = 0.05$ and (b) $\lambda = 0.01$.

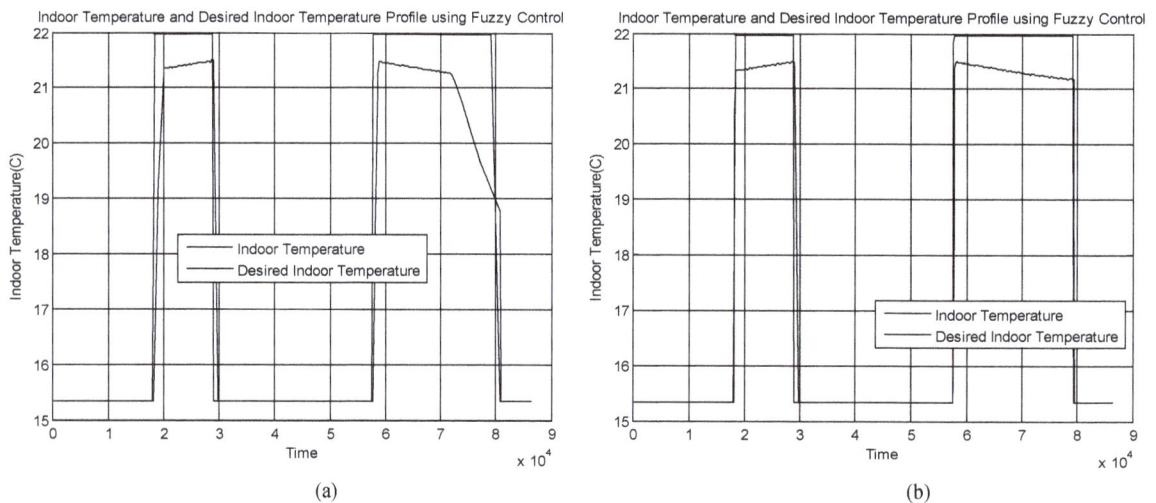

(a)   (b)

Figure 5. Comparison of actual and desired temperatures using a simple fuzzy controller for (a) $u_{max} = 0.1$ Kg/s and (b) $u_{max} = 0.5$ Kg/s.

Table 2. Comparison of energy consumption/day for Fuzzy, OSA and WOSA controllers.

OSA Controller		Fuzzy Controller		WOSA Controller $u_{max} = 0.1$ Kg/s	
$u_{max} = 0.1$ Kg/s	$u_{max} = 0.5$ Kg/s	$u_{max} = 0.1$ Kg/s	$u_{max} = 0.5$ Kg/s	$\lambda = 0.05$	$\lambda = 0.01$
2.944e + 8 J	2.962e + 8 J	2.010e + 8 J	2.018e + 8 J	1.944e + 8 J	1.962e + 8 J

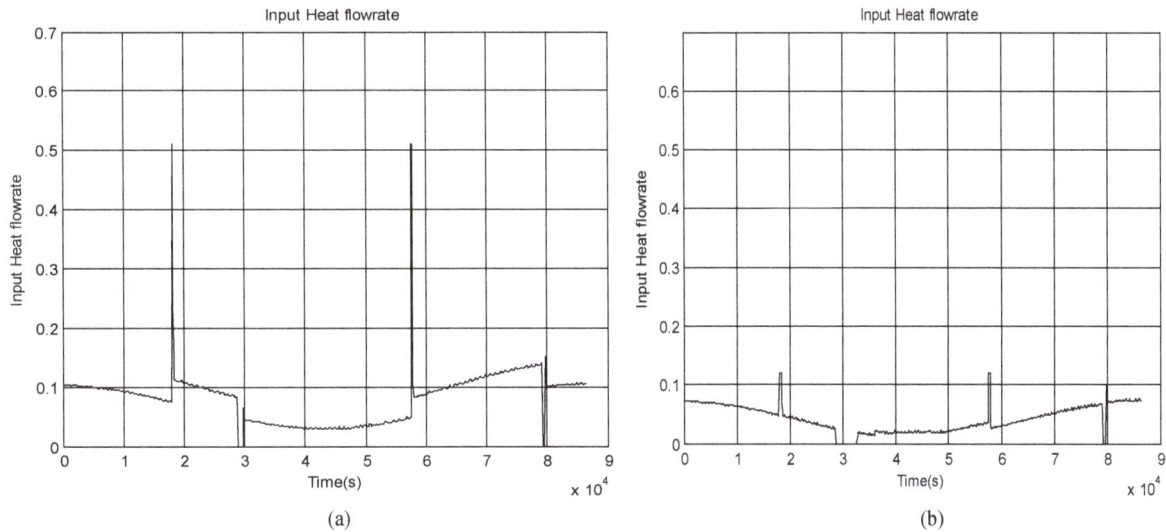

Figure 6. Behavior of fuzzy control signal for (a) $u_{max} = 0.1$ Kg/s and (b) $u_{max} = 0.5$ Kg/s.

HVAC systems are presented in this paper. A proof of global stability of the closed-loop system for the adaptive OSA controller is also presented. The performance of the proposed controllers is compared with that of a fuzzy-logic controller. The results of some simulation studies show that the adaptive OSA and WOSA controllers deliver better performance in terms of both tracking the desired temperature profile and reducing the overall energy consumption as compared to the Fuzzy controller. Further reduction of energy consumption by utilizing an optimized reference temperature profile is currently under investigation.

References

[1] Jimenez, M.J., Madsen, H. and Andersen, K.K. (2008) Identification of the Main Thermal Characteristics of Building Components Using Matlab. *Building and Environment*, **43**, 170-180. http://dx.doi.org/10.1016/j.buildenv.2006.10.030

[2] Paris, B., Eynard, J., Grieu, S., Talbert, T. and Polit, M. (2010) Heating Control Schemes for Energy Management in Buildings. *Energy and Buildings*, **42**, 1908-1917. http://dx.doi.org/10.1016/j.enbuild.2010.05.027

[3] Dounis, A.I. and Caraiscos, C. (2009) Advanced Control Systems Engineering for Energy and Comfort Management in a Building Environment—A Review. *Renewable and Sustainable Energy Reviews*, **13**, 1246-1261. http://dx.doi.org/10.1016/j.rser.2008.09.015

[4] Moon, J.W., Jung, S.K., Kim, Y. and Han, S.H. (2011) Comparative Study of Artificial Intelligence Based Building Thermal Control Methods—Application of Fuzzy, Adaptive Neuro-Fuzzy Inference System, and Artificial Neural Network. *Applied Thermal Engineering*, **31**, 2422-2429. http://dx.doi.org/10.1016/j.applthermaleng.2011.04.006

[5] Balan, R., Cooper, J., Chao, K.M., Stan, S. and Donca, R. (2011) Parameter Identification and Model Based Predictive Control of Temperature inside a House. *Energy and Building*, **43**, 748-758. http://dx.doi.org/10.1016/j.enbuild.2010.10.023

[6] Ma, Y., Borrelli, F., Hencey, B., Coffey, B., Bengea, S. and Haves, P. (2010) Model Predictive Control for the Operation of Building Cooling Systems. *IEEE Transactions on Control Systems Technology*, **20**, 796-803.

[7] Calvino, F., Gennusa, M.L., Morale, M., Rizzo, G. and Scaccianoce, G. (2010) Comparing Different Control Strategies for Indoor Thermal Comfort Aimed at the Evaluation of the Energy Cost of Quality of Building. *Applied Thermal Engineering*, **30**, 2386-2395. http://dx.doi.org/10.1016/j.applthermaleng.2010.06.008

[8]   Orosa, J.A. (2011) A New Modeling Methodology to Control HVAC Systems. *Expert Systems with Applications*, **38**, 4505-4513. http://dx.doi.org/10.1016/j.eswa.2010.09.124

[9]   Goodwin, G.C. and Sin, K.S. (1984) Adaptive Filtering Prediction and Control. Prentice-Hall, Englewood Cliffs.

[10] IBPT (2012) International Building Physics Toolbox in Simulink. http://www.ibpt.org/

[11] Antsaklis, P.J. and Michel, N.A. (2007) A Linear Systems Primer. Birkhauser (Springer), New York.

[12] Dochain, D. and Bastin, G. (1984) Adaptive Identification and Control Algorithms for Nonlinear Bacterial Growth Systems. *Automatica*, **20**, 621-634. http://dx.doi.org/10.1016/0005-1098(84)90012-8

[13] Goodwin, G.C., McInnis, B. and Long, R.S. (1982) Adaptive Control Algorithms for Waste Water Treatment and pH Neutralization. *Optimal control Applications and Methods*, **3**, 443-459.

[14] Fanger, P.O. (1972) Thermal Comfort: Analysis and Applications in Environmental Engineering. McGraw-Hill, New York.

9

Energy and Economic Growth, Is There a Connection? Energy Supply Threats Revisited*

Paul Ojeaga[1#], Deborah Odejimi[2], Emmanuel George[3], Dominic Azuh[3]

[1]Bergamo University, Bergamo, Italy
[2]Igbinedion University, Okada, Nigeria
[3]Covenant University, Ota, Nigeria
Email: [#]paul.ojeaga@unibg.it, ixerxes2001@yahoo.com

Abstract

The increased cost of accessing energy and the effects on economic growth (GDP) across regions is one of grave concern [1]. The Cost implication of energy supply often shapes regional energy policies across the globe. This paper presents an empirical investigation into the relationship between energy generation and economic growth, while also investigating probable threats to sustainable energy supply across regions. Energy generation was found to have some implications for economic growth across regions. It was found that hydro electric, renewable energy and nuclear generation sources were significantly driving growth across regions while coal and gas sources were not. This was particularly true since the cost of fossils was having strong cost implications, for overall energy generation cost in countries in regions due to overdependence on fossils. Generating sources were also found to have strong implications for sustained energy supply (energy security), renewable energy and gas generating sources that had the strongest effects on sustainable energy supply across regions. This was probably true since regions were focusing on new technologies in energy generation process, which are cheaper, cleaner and more sustaining, while still depending on gas plants due to the relative cost implications of maintaining gas plants compared to hydro and nuclear generating plants. The method of estimation used in the study is the seemingly unrelated regression estimation method.

Keywords

Energy Security, Energy Cost Reduction, Fossils, Growth, Renewable Energy

---

*This section introduces the trends in energy consumption, supply and generation across regions.
#Corresponding author.

1. Introduction

Lots of studies have already argued that energy generation has strong cost implication for private-sector driven growth. There also is an ongoing debate as to what exactly are the threats to sustainable energy supply [2], since the cost of generation can have strong implications for energy stakeholders and policy makers. While many studies have studied the impact of energy generation on growth, few studies have attempted to study the impact of different energy sources on growth, particularly as it affects regional growth as a main point of focus.

Issues of sustainability are also ones of paramount importance, and overdependence on fossils also means that susceptibility to failure in energy generation and supply risks is also increasing since issues of cost and political disputes e.g. the Russo-Ukrainian gas dispute of 2005/2006, can affect gas supplies across regions. While the use of fossils continues to have strong consequences for energy security, it is likely to have little or no effect on growth due to the cost implication of acquiring fossils for the energy generation process.

Environmental constraints, industrialization rate, domestic consumption characteristics and regional specific investment in domestic technology are possible determinants of energy availability across regions [1]. Global demands for energy are also on the increase in United Nations energy report 2012, making World energy consumption to have doubled by the year 2050.

Numerous literatures also continue to argue for diversification away from fossils due to overdependence [3], stating that diversification can lead to sustained supply and mitigate future risk of energy shortage attributable to cost related factors that affect gas supply availability. Other causal empirical studies, [4] also show that diversification is also on the increase in developed countries particularly the United States.

While fears of increased demands in the domestic energy markets of major exporters continue to increase, studies show that such demands are not likely to affect energy security on the short-run [5]-[7], since consuming countries are likely to shift to new exporters. In an attempt to study the cost reduction of the energy generation, [8] also argued extensively that inter-fuel substitution between oil and gas was of little significance compared to inter-fuel substitution between electricity and oil making fossils to have strong consequence for the electricity generation process.

The paper by [9] also studied the impact of portfolio diversity on cost for energy-importing countries and stated that consumer countries should hold portfolios free of cost risk associated with the hike in fossil fuel prices. The study by [10] also attempted to study the effects of cross-country energy policy on energy security from country-specific perspective, to energy vulnerability aversion, and they found that energy security had actually been affected by country-specific domestic consumption and reliance on specific sources for energy generation.

Many factors are known to affect growth across regions, while the neoclassical growth theory is based on the premise that technology is fixed across countries, and places strong emphasis on the importance of capital on growth, and the endogenous growth theory argues extensively on the importance that skill development can have on growth.

This paper examines the link between growth and energy generation across regions, and lots of factors have already been identified to affect growth across regions. They include: domestic energy consumption rate, industrialization rate, investment in domestic technology to boost energy generation, access to energy generation sources and other climatic concerns. The method of estimation used is the seemingly unrelated regression estimation method which has obvious advantages since it reduces the bias in two unrelated dependent variables of a simultaneous equation regression through the interaction of their errors with one another producing consistent estimates. The rest of the paper is divided into its scope and overview of study, empirical analysis and finally the concluding sections.

2. Overview of Study

2.1. Scope and Objective of Study

The scope of the paper presents empirical evidence on the energy generation and growth and revisits threats to sustained energy supply across regions previous addressed by [11]. The objectives of the study include:
   1) Does energy generation affect growth across regions?
   2) What energy generation sources are relevant to driving growth across regions?
   3) Do threats identified as risk to energy supply matter in increasing supply risks across regions?

2.2. Stylized Facts on Energy Security and Growth across Regions

In this section we present the stylized facts on energy security and growth across regions. World energy generation capability is on the increase, with output energy in Asia likely to surpass total generation in Europe and North America by 2040 (see **Table 1**) that growth across regions is on the increase except for Africa (see **Figure 1** id 3), while growth is on the increase in North America, European, Union, Latin America and South East Asia. Most countries in Africa remain poor despite sustained commodities high prices in the global market. In many countries that are experiencing growth in Africa the growth is also not inclusive.

Table 1. World installed generating capacity by region and country, 2010-2040.

OECD	Projections							Yearly % Changes
	2010	2015	2020	2025	2030	2035	2040	
OECD Americas	**1248**	**1316**	**1324**	**1379**	**1456**	**1546**	**1669**	**1.0**
United States[a]	1033	1080	1068	1098	1147	1206	1293	0.8
Canada	137	144	152	163	174	185	198	1.2
Mexico/Chile	78	93	104	118	135	155	177	2.8
OECD Europe	**946**	**1028**	**1096**	**1133**	**1159**	**1185**	**1211**	**0.8**
OECD Asia	**441**	**444**	**473**	**489**	**501**	**516**	**524**	**0.6**
Japan	287	275	293	300	304	309	306	0.2
South Korea	85	93	100	107	114	122	130	1.5
Australia/New Zealand	69	76	81	83	83	85	87	0.8
Total OECD	**2635**	**2788**	**2894**	**3002**	**3116**	**3247**	**3403**	**0.9**
Non-OECD								
Non-OECD Europe and Eurasia	**408**	**421**	**455**	**480**	**508**	**538**	**563**	**1.1**
Russia	229	239	264	282	299	315	325	1.2
Other	179	182	191	198	209	223	239	1.0
Non-OECD Asia	**1452**	**1820**	**2188**	**2479**	**2772**	**3057**	**3277**	**2.8**
China	988	1301	1589	1804	2007	2176	2265	2.8
India	208	241	285	327	376	440	510	3.0
Other	256	278	314	347	390	441	502	2.3
Middle East	**185**	**197**	**216**	**233**	**247**	**267**	**280**	**1.4**
Africa	**134**	**147**	**164**	**184**	**211**	**244**	**283**	**2.5**
Central and South America	**247**	**279**	**304**	**329**	**362**	**400**	**447**	**2.0**
Brazil	114	137	152	169	191	221	256	2.8
Other	134	142	152	160	171	179	191	1.2
Total Non-OECD	**2426**	**2864**	**3327**	**3705**	**4099**	**4505**	**4850**	**2.3**
Total World	**5061**	**5652**	**6221**	**6707**	**7214**	**7752**	**8254**	**1.6**

[a]Includes the 50 states and the District of Columbia. Note: Totals may not equal sum of components due to independent rounding. Sources: History: Derived from U.S. Energy Information Administration (EIA), International Energy Statistics database (as of November 2012), www.eia.gov/ies. Projections: EIA, Annual Energy Outlook 2013, DOE/EIA-0383(2013) (Washington, DC: April 2013); AEO2013 National Energy Modeling System, run REF2013.D102312A, www.eia.gov/aeo; and World Energy Projection System Plus (2013).

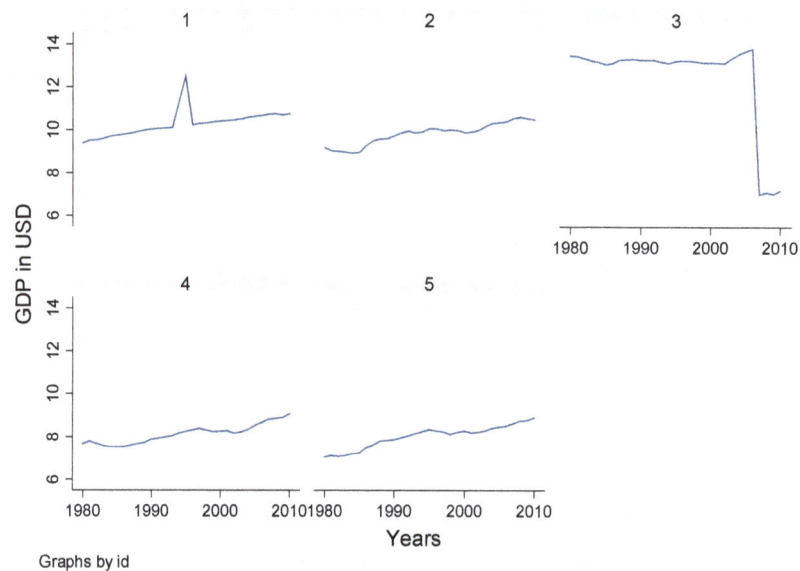

Note: The graphs above show trends for North America, Europe, Africa, Latin America and South East Asia respectively.

Figure 1. Regional GDP trends.

Despite been endowed with enormous natural resources Africa also remains plagued with poor governance and weak institutions making many policies not to have any effect on growth and economic development in the region, [12]-[14]. North America still remains the most productive of all regions this is attributable to high productive of it level force which remains the most productive due to high skill and technological endowments associated with the region [14]. Other regions which include Europe, Latin America and South East Asia are also experiencing significant growth.

Trends also show that energy security is also low for Africa (see **Figure 2**), depicting poor implementation of the Kyoto Protocol as well as an under developed energy sector plagued with high energy supply and distribution disruptions. Issues associated with the cost implications of developing energy plants is also a problem in many African countries with poor income. The Doha round of talks breakdown will also have strong implications for energy security since major energy consumers e.g. United States and Canada pulled out are not likely to commit to emission reduction targets set by regulatory agencies. Therefore it is not expected that regions are likely to be alive to the negative effects associated with poor energy consumption methods currently on grounds which can lead to potential supply problems in the future.

The rapid industrialization in Latin America, South East Asia and also in some emerging African countries starting in the early 2000s, see id 3, 4 and 5 respectively in **Figure 8**, also means that the competition for the world resources is on the increase despite the slowdown in the industrialization development of the highly developed countries in Europe and North America (see id 1 and 2 in **Figure 3**). Investment in domestic technology in regions is also ongoing with a steady rate of investment in Europe and North America and continuous improvement for Latin America and Africa. North America particularly the United States and Canada have some of the largest number of Wind generating plants in the World after China [15]. Asia is presently experiencing a slowdown from the massive investment of the 1990s in generation technology, but still maintaining steady investment in the development of improved generation sources.

North America has the most diversified energy sector with the United States having the highest number of wind farms and hydro power stations in the world (see **Figure 2** where energy security is the measure of how diversified the regional generation process is). Europe is also gradually disengaging from the use of nuclear plants in energy generation and introducing renewable energy technology in the energy generation process although it is still vulnerable to strong dependence on Gas production sources [1].

Hydro production capabilities utilization is still reasonably high for North America, Europe, Africa and Latin America (see **Figure 3**). South East Asia is actually experiencing reduced dependence on hydro generation due

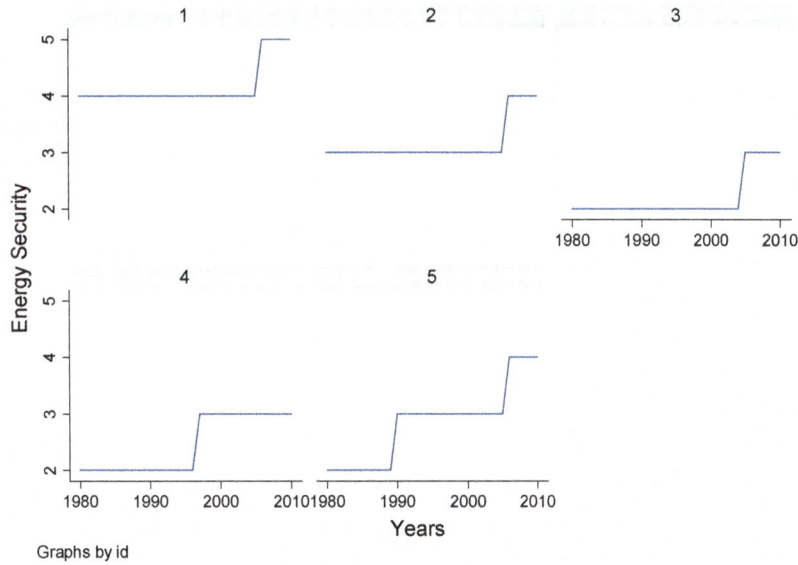

Note: The graphs above show trends for North America, Europe, Africa, Latin America and South East Asia respectively.

Figure 2. Energy security trends across regions.

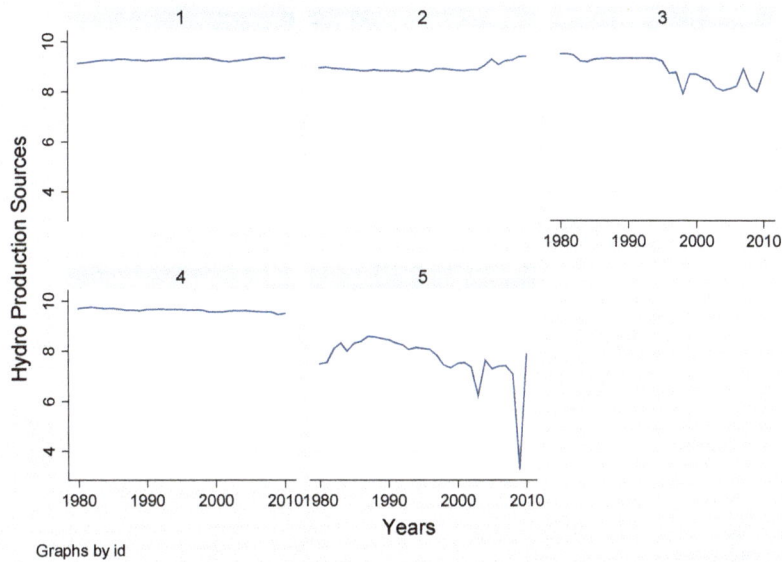

Note: The graphs above show trends for North America, Europe, Africa, Latin America and South East Asia respectively.

Figure 3. Regional hydro generation.

to probably poor natural sources for developing hydro generation plant capabilities. The use of coal in energy generation is also on the increase for all regions except North America and Europe where declines in their use are noticeable (see **Figure 4**). This is probably due to the advent of alternative means of generation that are cleaner making these highly developed regions to lack further incentives to continue developing more of such plants for future energy use.

Dependence on nuclear generating plants is also on the decrease in all regions except in Africa where only minimal increases were recorded; this is attributable to complexities associated with nuclear waste disposal, cost of maintenance and development and finally the high risk associated with operating such plants, making regions not to have sufficient incentive to develop such generating capacities (see **Figure 5**). Uses of renewable sources were also on the increase except for parts of Asia and Latin America (see **Figure 6**).

Reliance on gas production sources are also on the increase for all regions except for Africa, this is attributable to the relative ease of development of gas plants and access to gas supplies to power such plants. The use of gas plants in Africa has not experienced commensurate increase compared to other regions due to issues of poor technology and the cost implications of developing such plants since such technologies are often obtained

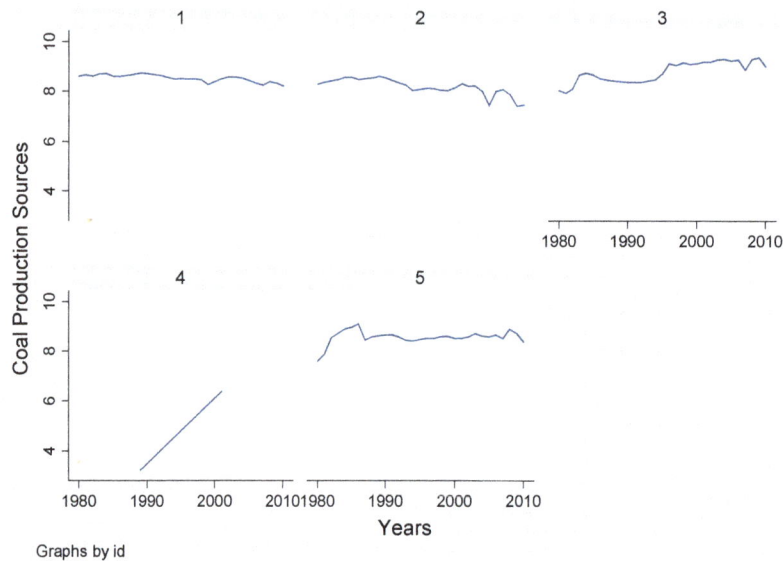

Note: The graphs above show trends for North America, Europe, Africa, Latin America and South East Asia respectively.

Figure 4. Regional coal production.

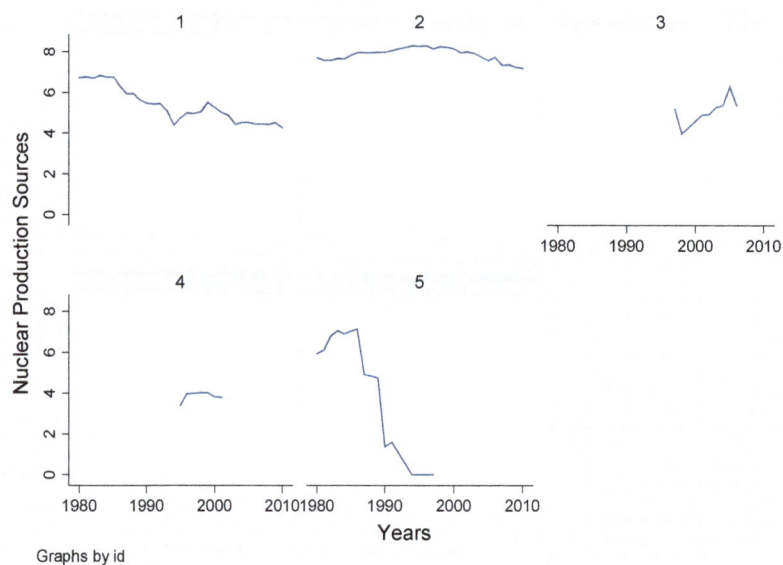

Note: The graphs above show trends for North America, Europe, Africa, Latin America and South East Asia respectively.

Figure 5. Regional nuclear production.

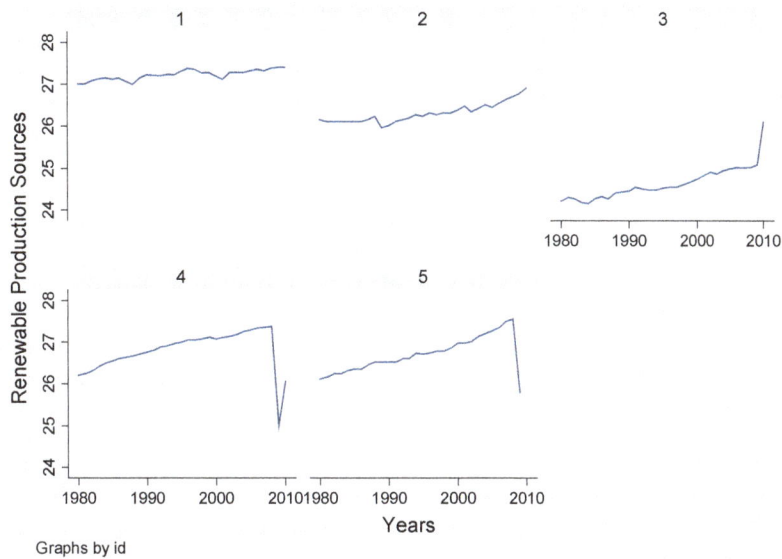

Note: The graphs above show trends for North America, Europe, Africa, Latin America and South East Asia respectively.

Figure 6. Renewable energy production.

overseas (see **Figure 7**).

Energy use in general across regions is on the increase making regions to be vulnerable. Population growth and industrial development in regions continue to exert strain on current generation infrastructure making countries in regions to be constantly engaged in development of more plants and use of cheaper and alternative methods in the generation process (see **Figure 8**). Finally an increase in domestic demand is driving generation and development of more energy plants (see **Figure 9**).

3. Empirical Analysis and Results

3.1. Empirical Analysis

In this section we present the empirical details used in the study, data for the study is obtained from data market of Iceland for the period of 1980 to 2010, 31 years, for five regions which include for North America, Europe, Africa, Latin America and South East Asia respectively. The dependent variables include energy security (ENSEC) which we measure using score values assigned to regions, based on the level of diversification and infrastructure in renewable energy sources in regions with North America particularly the United States having stronger capabilities towards averting energy interruptions, and economic growth which is the aggregate regional gross domestic product in constant US dollars (USD). We take the logarithm of GDP (Log GDP) due to its noisiness.

Other explanatory variables include cost of accessing energy resources, electricity output generation by source was captured from each generation source such as nuclear plants (Nuclear Prod.), renewable energy sources (particularly wind energy and biogas productions) (Renew Prod. coal powered electric plants (Coal Prod.), gas driven turbines (Gas Prod.) and hydroelectric production (Gas Prod.) which even though classified as renewable was separated from what was called renewable in this study due to strong dependence on hydroelectric generating plants, all measured in kilowatts hours (KWH), environmental constraints (Env. Const.), measured using number of dams, regional size and average regional temperatures that are likely to affect electricity transmissions and consumption particularly for temperate regions were used to generate an index for environmental constraint see [16] for more information on regression residual index generation). Countries, firms and individuals are likely to have fixed budgets, making budget constraints to be an issue for access consumption. For countries the effects will be two fold it will affect the cost of access limiting the amount accessible and it will have negative effect on growth shrinking individual countries in regions budgetary allocation due to the

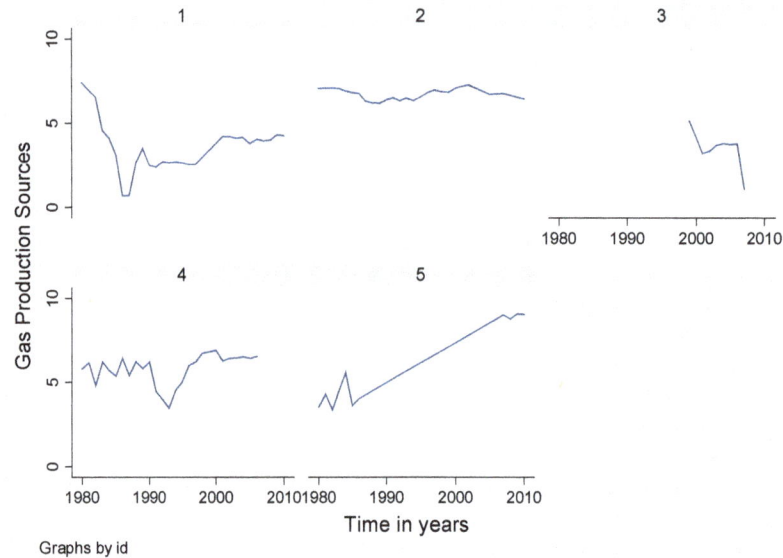

Note: The graphs above show trends for North America, Europe, Africa, Latin America and South East Asia respectively.

Figure 7. Regional gas production.

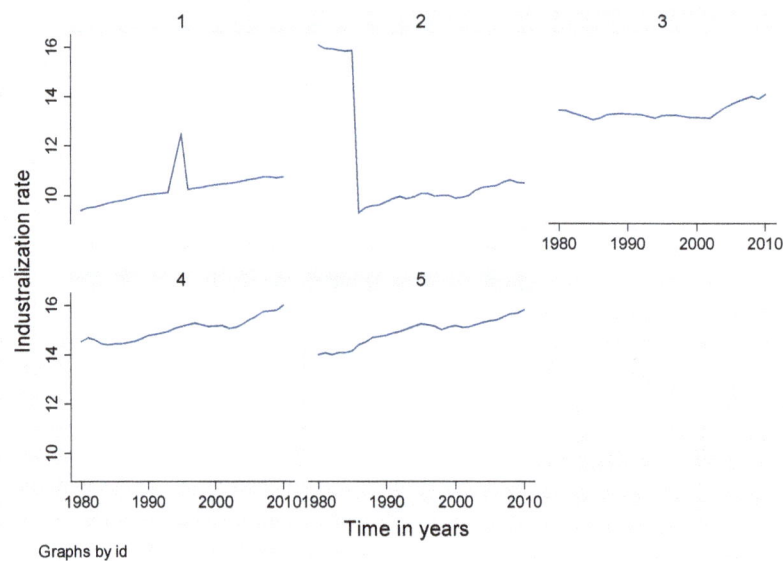

Note: The graphs above show trends for North America, Europe, Africa, Latin America and South East Asia respectively.

Figure 8. Regional industrialization trends.

capital intensive nature of building power plants and accessing resources to power them. This is measured using GDP per capital divided by percentage inflation to deflating for inflation to obtain real wages. Regional specific investment in domestic innovation (Inv. Dom.) was also measured using total investment in research and development in regions in constant US dollars. While regional industrialization rate was regional specific logarithm of GDP in constant USD which will be high for highly industrialized countries. Energy consumption rate (Ene Cons.) was measured using total domestic consumption for countries in regions in Kilowatt hour and finally regional specific energy policy was measured using score values for regional specific participation and implemen-

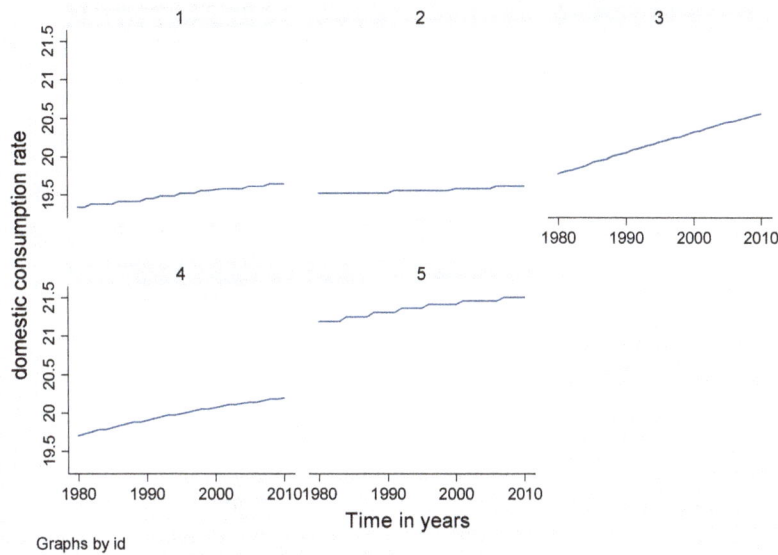

Note: The graphs above show trends for North America, Europe, Africa, Latin America and South East Asia respectively.

Figure 9. Domestic energy consumption across regions.

tation of the Kyoto protocol starting from 1998 when the first inter government panels were set up to 2010 when commitment towards emission reduction and implementation plans were emphasized also using score values of 1 to 3 depending on regional level of implementation and finally using consumption patterns in the pre Kyoto protocol years.

3.2. Model Specification and Equations

Energy Diversification will be the number of existing dependable generation sources, while energy policy across regions will be a function of energy diversification and countries across regions signatory and to the Kyoto protocol expressed below as:

$$\text{Policy} = \text{Energy Diversification} + \text{Kyoto Protocol Signatory} \tag{1}$$

The reason for this is that policy in countries across regions will be shaped by the need to stem supply interruption through diversifying the generation process as well as reducing greenhouse gases which is one of the aims of the Kyoto accord, which could lead to countries exploring cleaner and possibly cheaper methods of energy generation such as the use of renewable energy sources in the generation process.

Environmental constraints such as the availability of natural resources for the generation process is also taken into consideration, other issues such as size of regions since transmission of energy through long distances is costly due to losses that accompany such circumstances are also a matter of concern and finally regional temperature level are also an issue of concern since temperate regions are likely to need sustained heating supply during cold winter conditions which could be costly and tropical regions are likely to harness cheap solar source if not now but in the future. This allows us to state that environmental constraints to the generation process will be a function of energy resource, regional size and differences in regional temperature levels:

$$\text{Env. Const.} = \text{Available Energy Resource} + \text{Regional Size} + \text{Temperature Levels} \tag{2}$$

Energy consumption will also be a function of three factors income which will affect how consumers can afford available energy, available supply of energy across regions and aggregate demand across regions which will be a function of population density expressed below as:

$$\text{Energy Consp.} = \text{Income} + \text{Energy Supply} + \text{Population Density} \tag{3}$$

Domestic technology will be a function of current technology and investment in technology which will affect technology focus overtime expressed below in Equation (4):

$$\text{Domestic Tech} = \text{Current Tech} + \text{Investment in Technology} \qquad (4)$$

Energy generation will be a function of diversification depicted by number of dependable generating sources across regions, environmental constraints, available technology as well as domestic and industrial demands for energy expressed below as:

$$\text{Energy Generation} = \text{Diversification} + \text{Env. Const.} + \text{Tech} + \text{Energy Consp.} \qquad (5)$$

Energy security will be a function of the energy generation process and energy policy across regions which will depict how regions efficiently deploy their resources to provide cheap and sustainable energy across regions:

$$\text{Energy Security} = (\text{Diversification} + \text{Env. Const.} + \text{Tech} + \text{Energy Consp.}) \text{ Policy} \qquad (6)$$

Generation (GEN) and energy security can be written as a function of diversified energy sources across regions expressed as GEN f (SOURCES, DOM. TECH., ENV. CONST. and ENE. CONS.) and ENE SEC f (GEN, POLICY)

Energy security is the regarded as the access to uninterruptible energy supply in regions EU Green Paper 2001, the method of identification is based on the fact that energy security will be a function of diversified generation sources and other factors such as environmental constraints that affect that affect the energy generation and supply process, consumption demands both domestic and industrial, availability of domestic technology since it is easier and cheaper to access technology domestically than oversea, regional specific energy policy which will show how regions are deploying their resources towards achieving cheaper, better, cleaner and more sustainable methods of energy supply and generation and finally access to capital which in this case depicts the cost of acquiring capital across regions.

$$\text{Growth}_{it} = A K_{it}^{\alpha} L_{it}^{1-\alpha} \qquad (7)$$

$$\text{Growth}_{it} = A K_{it}^{\alpha} L_{it}^{1-\alpha}. \text{ Energy Generation}_{it} \qquad (8)$$

Growth will also be a function of cost of labor, cost of access to capital across regions, the fixed level of technology across regions, energy generation, environmental constraints which often increase the cost of energy generation and supply since it can increase the cost of production particularly for the private sector across regions, and finally domestic and industrial demand for energy across regions since this can affect cost of energy at delivery which can have strong effects on final prices of goods and affect income across regions expressed below as GROWTH f (TECH, CAPITAL, LABOUR and GEN).

The method of estimation used in the study is the seemingly unrelated estimation method, based on the assumption that solving two unrelated equations simultaneously will reduce the bias of the error terms $u_{1i,t}$ and $u_{2i,t}$ as shown in the model specification below,

$$\text{ENSEC}_{i,t} = \alpha_0 + \alpha_1 GS_{i,t} + \alpha_2 X_{i,t} + u_{1i,t} \qquad (9)$$

$$\text{GROWTH}_{i,t} = \lambda_0 + \lambda_1 \text{LABOUR}_{i,t} + \lambda_2 \text{CAPITAL}_{i,t} + \lambda_3 \text{TECH}_{i,t} + \lambda_4 X_{i,t} + u_{2i,t} \qquad (10)$$

through their interaction in a simultaneous regression model. In this case we study the dynamics between energy supply (energy security) potential threats to such supply such as number of generation sources, environmental factors that affect such supply, availability of technology, consumption rate across regions etc. and also the possible effects of energy generation on economic growth through the investigation of what generation sources were promoting growth across regions using the same set of exogenous variables. The variable year is also included to control for annual differences in energy supply and generation that may affect energy security and growth across regions. We do not believe the model will suffer from mis-specification since it identifies all major factors that affect energy security as well growth across regions.

3.3. Results

In this section we present the results for the energy security and growth equations respectively in **Table 2** and **Table 3** below. Renewable and gas production sources were found to be improving energy security in a significant manner. Investment in domestic technology and energy policy across regions were also found to affect energy-security in a positive significant manner reducing vulnerability to generation and supply interruption across regions.

Table 2. Regression showing the potential threats to energy supply.

Variables	(1) Energy Security	(2) Energy Security	(3) Energy Security	(4) Energy Security	(5) Energy Security
Cost of Energy	−1.21*** (2.45)	−1.33*** (2.50)	−3.57 (2.77)	−1.20*** (2.39)	−1.18*** (2.45)
Energy Consumption	−2.71*** (8.10)	−2.13*** (7.90)	−2.24*** (6.83)	−2.49*** (8.42)	−2.78*** (7.75)
Dom. Innovation	0.0002*** (4.32)	0.0002*** (4.07)	0.0001*** (4.20)	0.0003*** (4.54)	0.0003*** (4.47)
Env. Constraint	−7.93*** (1.69)	−8.12*** (1.67)	−4.18** (1.68)	−8.17*** (1.72)	−7.48*** (1.73)
Indust. Rate	−8.57*** (2.52)	−8.03*** (2.46)	−8.40*** (2.30)	−1.26*** (3.91)	−7.53*** (2.76)
Energy Policy	0.93*** (0.14)	0.92*** (0.13)	0.89*** (0.12)	0.94*** (0.16)	0.90*** (0.13)
Hydro Production	4.79 (1.06)				
Coal Production		3.11 (1.89)			
Renewable Prod.			0.01*** (0.01)		
Gas Production				0.0001** (4.74)	
Nuclear Prod.					7.65 (7.70)
Year Effect	Yes	Yes	Yes	Yes	Yes
Observations	112	113	112	88	113
R-Squared	0.71	0.72	0.77	0.78	0.72

Note: The results presented above show the effect of different energy sources on energy security across regions. Renewable energy has strong effects on energy security and overdependence on gas was found to still be present across regions. Standard errors are in parentheses ***$p < 0.01$, **$p < 0.05$, *$p < 0.1$.

Table 3. Regression showing the impact of various generating sources on growth.

Variables	(1) Log of GDP	(2) Log of GDP	(3) Log of GDP	(4) Log of GDP	(5) Log of GDP
Average Wages	8.00*** (2.80)	8.13*** (3.15)	8.32*** (3.92)	7.72*** (2.88)	8.02*** (3.04)
Energy Consumption	−1.49 (9.25)	−5.90 (9.93)	5.99 (9.64)	−1.34 (1.02)	−3.97 (9.61)
Domestic Innovation	0.0001*** (4.94)	0.0002*** (5.11)	0.0002*** (5.93)	0.0001* (5.48)	0.0003*** (5.55)
Env. Constraint	−0.0002*** (1.93)	−0.0002*** (2.10)	−0.0002*** (2.37)	−0.0002*** (2.07)	−0.0002*** (2.15)
Indust. Rate	−1.82*** (2.88)	−2.33*** (3.10)	−2.16*** (3.25)	−2.30*** (4.72)	−1.67*** (3.42)
Energy Policy	0.77*** (0.15)	0.78*** (0.17)	0.73*** (0.17)	0.40** (0.20)	0.62*** (0.17)
Hydro Production	6.65*** (1.21)				
Coal Production		−6.78*** (2.38)			
Renewable Prod.			0.01* (0.001)		
Gas Production				7.65 (5.72)	
Nuclear Prod.					0.0003*** (9.55)
Year Effect	Yes	Yes	Yes	Yes	Yes
Observations	112	113	112	88	113
R-Squared	0.92	0.91	0.91	0.95	0.91

Note: The result above show the effect of generating sources on growth across regions coal production source was having a negative effect on growth. Renewable energy and nuclear production sources promoting growth Standard errors are in parentheses ***$p < 0.01$, **$p < 0.05$, *$p < 0.1$.

Cost of access to energy, energy consumption, environmental constraint and industrialization rate were having a negative significant effect on energy security increasing the risk to generation and supply interruption across regions (see **Table 2**).

Renewable energy, hydro energy and Nuclear energy sources were found to be promoting economic growth across regions. While industrialization rate have peaked for the developed countries continuous investment in domestic technology and energy policy were found to improve economic growth across regions. Coal production source was having a negative effect, on growth in countries, across regions. Environmental constraint was having a negative effect on economic growth depicting that increased cost of accessing energy resources were still an issue in countries across regions. The variable year was also significant showing that energy generation was having an effect on energy security and growth, the differences in fluctuations in growth and energy security in years were likely not be responsible for changes in energy security.

All the objectives of the study were realized, the question if energy generation was affecting growth across regions was answered although coal production sources were having negative effect on growth, renewable and nuclear source were positively driving growth across regions. Renewable energy sources were found to be most relevant to driving growth across regions, over-dependence on nuclear sources were also having positive implications for growth this will most likely affect developed countries that utilize this source of generation. Finally, some threats were found to negatively affect energy security in general they included the cost of accessing energy, environmental issues associated with the generation process, growing domestic consumption demands and industrial demands for energy.

4. Discussion and Conclusions

In this section we discuss the implications of our findings. The world is heading towards another challenging era since global energy demands are on the increase. Threats were found to matter in the energy production and access process. Some threats identified, negatively affect energy security in general including the cost of accessing energy, environmental issues associated with the generation process, growing domestic consumption demands and industrial demands for energy.

It was found that energy generating sources were improving energy security in regions with renewable and gas energy sources that had positive significant effects on mitigating energy generation disruptions across regions. The implication of this finding was that the issue of overdependence on fossil supply sources was probably an issue regions have to take strongly. While gas production were improving, energy security regions were still susceptible to environmental risks such as pollution from green house (gases) and gas supply disruption associated with cost implications and other political factors such as gas channel disputes that could affect gas supply from producing countries to destination countries. Energy generating sources were also found to have implicative consequences for growth across regions with nuclear and renewable energy sources having positive effects on growth. This result was however not same for coal production sources as they were found to have growth reduction effects. This is probably attributable to environmental consequences of using coal leading to reduced use of coal in regional energy generation capabilities in general.

Finally regions were found to depend more and more on renewable energy sources, and this would be particularly true for developed countries and could also have strong implications for developing countries with low income since renewable energy sources are likely to be cheaper and a more reliable means to generate energy with minimal environmental pollution and degrading consequences, in countries across regions. Improving energy security through diversifying generation could also mean improvements in economic growth particularly for the private sector since it could improve private sector-led growth and aid industrialization efforts in countries.

The study supports past findings by [9] who stated that diversification were necessary to improve energy security in general through holding risk-free portfolios to stem risk associated with hikes in fossil prices and the study by [10] who argued that energy policy were important in averting energy crisis since resources for generation were often scarce leading to the competition for the available scarce resources. The findings have strong policy implications for stake holders in the energy industry and for governments, and renewable energy sources are probably a way to go in the future and have far-reaching effects on growth evidence which are evident in their environmental friendliness, sustained use and relative cheapness having obvious implications for developing countries in particular.

Acknowledgements

"P. Ojeaga" thanks Bergamo University Italy and Prof. Annalisa Cristini specifically, for funds in obtaining the data, for this work during his Ph.D. research. He also expresses "Thanks" to the Chief Librarian in Watertown MA (Boston) for allowing him to use his favorite corner while collating the data for this study.

References

[1] International Energy Agency (IEA) (2001) Report 2001.

[2] European Commission (2000) Towards a European Strategy for the Security of Energy Supply. Green Paper, COM769, Brussels.

[3] Cohen, G., Joutz, F. and Loungani, P., (2011) Measuring Energy Security: Trends in the Diversification of Oil and Natural Gas Supplies. IMF Working Paper No. 11/39, Washington, DC.

[4] Bryce, R. (2008) The Dangerous Delusions of "Energy Independence". Public Affairs, New York.

[5] Le Coq, C. and Paltseva, E. (2008) Common Energy Policy in the EU: The Moral Hazard of the Security of External Supply. SIEPS Report 2008: 1, Stockholm, Sweden

[6] Neumann, A. (2004) Security of Supply in Liberalised European Gas Markets. Diploma Thesis, European University Viadrina, Viadrina.

[7] Neumann, A. (2007) How to Measure Security of Supply? Mimeo, Dresden University of Technology, Dresden.

[8] Annual Energy Outlook (2009) Energy Information Administration Office of Integrated Analysis and Forecasting. U.S. Department of Energy.

[9] Awerbuch, S., Jansen, J.C. and de Vries, H.J. (year) Demonstrating and Building Capacity for Portfolio-Based Energy Planning in Developing Countries. Interim Report Submitted to The Renewable Energy & Energy Efficiency.

[10] Knox-Hayes, J., Brown, M.A., Sovacool, B.K. and Wang, Y. (2013) Understanding Attitudes toward Energy Security: Results of a Cross-National Survey. *Global Environmental Change*, **23**, 609-622. http://dx.doi.org/10.1016/j.gloenvcha.2013.02.003

[11] Ojeaga, P., Odejimi, D. and Alege, P.O. (2013) Rethinking Regional Energy Policy: Towards Averting AnotherEnergy Crisis. Do Threats Matter in The Supply and Generation Process? *Journal of Energy Technologies and Policy*, **4**, 1-17.

[12] Alexeev, M. and Conrad, R. (2009) The Elusive Curse of Oil. *The Review of Economics and Statistics*, **91**, 586-598. http://dx.doi.org/10.1162/rest.91.3.586

[13] Bhattacharyya, S. (2009) Root Causes of African Underdevelopment. *Journal of African Economies*, **14**, 745-780.

[14] Acemoglu, D., Johnson, S. and Robinson, J. (2001) The Colonial Origins of Comparative Development: An Empirical Investigation. *American Economic Review*, **91**, 1369-1401.

[15] Renewable Global Status Report REN21 (2012) Renewable Global Status Report 2006-2012.

[16] Burnside, C. and Dollar, D. (2000) Aid, Policies and Growth. *American Economic Review*, **90**, 847-868.

10

The Identification of Peak Period Impacts When a TMY Weather File Is Used in Building Energy Use Simulation

Jay Zarnikau[1,2*], Shuangshuang Zhu[1]

[1]Frontier Associates LLC, Austin, USA
[2]LBJ School of Public Affairs and Division of Statistics and Scientific Computing, The University of Texas at Austin, Austin, USA
Email: [*]jayz@frontierassoc.com

Abstract

When typical meteorological year (TMY) data are used as an input to simulate the energy used in a building, it is not clear which hours in the weather data file might correspond to an electric or natural gas utility's peak demand. Yet, the determination of peak demand impacts is important in utility resource planning exercises and in determining the value of demand-side management (DSM) actions. We propose a formal probability-based method to estimate the summer and winter peak demand reduction from an energy efficiency measure when TMY data and model simulations are used to estimate peak impacts. In the estimation of winter peak demand impacts from some example energy efficiency measures in Texas, our proposed method performs far better than two alternatives. In the estimation of summer peak demand impacts, our proposed method provides very reasonable results which are very similar to those obtained from the Heat Wave approach adopted in California.

Keywords

Peak Demand Reduction; Energy Efficiency Impact Analysis; Building Energy Use Simulation

1. Introduction

It is not always clear how weather-sensitive energy efficiency measures will perform at the exact hour(s) of the utility's annual summer or winter system peak. Often, building energy use simulation models are used to obtain

---
[*]Corresponding author.

8760 hourly impact estimates for the change in load associated with the efficiency measure based on typical meteorological year (TMY) data. TMY data contain actual months of weather data from different past years. Consequently, the TMY year does not coincide with any actual year and thus cannot be matched against actual demand or load data for a utility system or market. The hours associated with the most extreme temperatures in a TMY file may not necessarily correspond with a peak in demand in a utility system. Other factors, such as day of the week and the hour within the day, may also play a role. The challenge is to determine which of the 8760 hourly values obtained from a building energy use simulation model to select to represent the demand reduction at the time of the utility's system peak.

This topic is of great importance to utility system planners. Electric and natural gas utility systems are constructed largely to meet peak demand. Thus, the impact or performance of an energy efficiency measure during peak periods is of keen interest. Energy efficiency measures or demand side management (DSM) programs are valued, in part, based upon the generation and transmission costs which they could potentially displace [1]. Thus, the potential for an energy efficiency measure to reduce demand during the system's peak affects the value of measures and programs.

Various utility regulatory commissions in the US provide specific instructions for utilities and energy efficiency program administrators to follow when selecting the hour(s) associated with peak demand impacts. In California's DEER database, "the demand savings due to an energy efficiency measure is calculated as the average reduction in energy use over a defined nine-hour demand period" [2]. These nine hours correspond with 2 pm to 5 pm during 3-day heat waves. The Mid-Atlantic Technical Reference Manual, used in Maryland, Delaware, and DC, states: "The primary way is to estimate peak savings during the most typical peak hour (assumed here to be 5 pm) on days during which system peak demand typically occurs (i.e., the hottest summer weekdays). This is most indicative of actual peak benefits." [3] The New York Public Service Commission instructs: "Program Administrators (PAs) should calculate coincident peak demand savings based on the hottest summer non-holiday weekday during the hour ending at 5 pm." [4] Wisconsin's utility regulatory agency requires that peak demand reduction for weather-sensitive efficiency measures be based on average design-day conditions [5]. In Illinois, Colorado, New Jersey, and Maine, coincidence factors are used to estimate peak demand reduction based the impacts of a weather-sensitive efficiency measure on annual energy consumption [6]-[9]. For energy efficiency measures which are not weather-sensitive, a number of states find it acceptable to average the expected energy impacts of the measure over a large number of hours within some "peak period". We are unaware of any regulatory authority having adopted a formal probabilistic approach to estimating the impacts of energy efficiency measures upon the peak demand of a utility or market.

A formal probabilistic approach is attractive because the system peak of a utility or a market may not necessarily coincide exactly with the hottest summer day, historical temporal patterns, design-day conditions, or heat waves. For example, an extremely hot summer temperature reading may not necessarily lead to a summer peak, if the extreme temperature occurs on a weekend (when energy use in the commercial or business sector may be lower) or early in the afternoon (before the occurrence of an after-work peak in household energy use). While extreme temperatures may be the most important determinant of system peak demand, various patterns in energy usage (as might be reflected by the time of day and the month of the year), and other factors may play a role as well. A probabilistic approach can be used to quantify how various factors may contribute to the establishment of a peak in system demand.

2. Proposed Approach

Our proposed approach to matching a seasonal peak on a utility system with a TMY data file involves the following steps:
- Establish the number of hours to be included in the set of peak hours to be predicted each year and season (e.g., summer and winter).
- Use a logistic regression model and hourly data for a number of historical years to estimate the relationship between setting a peak hour and a set of explanatory variables, including a temperature variable and dummy variables representing the time-of-day and month-of-year.
- Use the estimated relationships to assign marginal probabilities to changes in the explanatory variables.
- With the estimated relationships, calculate the probability of setting a peak hour based on TMY weather data.
- Find and average the savings (i.e., the difference between a base and change case) from the outputs of a

building energy use simulation model that used the same TMY data file which corresponds to the same hours.

Although system planners often use a single hour or 15-minute interval to measure peak demand, predicting a larger set of peak hours tends to be more practical in the first step. Building energy use simulation models have stochastic algorithms. So if a single pair of model runs (*i.e.*, a base case and a change case) is used to calculate hourly savings, the predicted savings may be biased for any single hour. So, either multiple model runs must be used to average the estimated hourly savings, or a broader definition of peak (*i.e.*, peak hours) must be used. Further, in the analysis of the cost-effectiveness of energy efficiency measures and programs, the demand reduction tends to be valued based on the capital cost of a combustion turbine which normally has an expected annual runtime of 10 to 40 hours. Thus an analysis of the cost-effectiveness of energy efficiency measures and programs may benefit from knowledge of the impacts over a set of hours. Finally, estimating the probability of setting a set of peak hours is much easier than estimating a single peak hour or interval per year with a logistic model. For example, if six years of historical data are used and thus Y = 1 on only six instances, more advanced techniques would be required in the estimation (e.g., the use of a prior distribution and Bayesian estimation techniques) than those discussed here. For these reasons, a set of 20 peak hours is used in the examples presented here.

Note that the second step ignores many other very important factors that might affect the timing of the peak, including actions by industrial energy consumers and load-serving entities to respond to wholesale market price spikes[1]. The day of the week is also not considered. However, the inclusion of other variables would prevent the application of this approach when only a TMY weather file and a building energy use simulation model are used to calculate the peak demand reduction associated with an efficiency measure. TMY data are pieced-together from recorded weather during numerous previous years to create a typical year with typical fluctuations. Since the TMY data do not represent weather data from any single "real" year, there would be no way of matching "real" energy price data, the day of the week, or other variables to the fabricated weather data.

Marginal probabilities can be obtained by estimating a logistic regression or logit model [11]. Most statistical software packages can convert the results from a logit model into probabilities [12] [13].

In the final step, either a simple average or a probability-weighted average (with the weights based on the probability of the seasonal peak being set in a particular hour in the TMY data file) could be used to estimate peak demand reduction among those hours within the set of peak demand hours.

3. An Example Determination of Peak Hours

An example is illustrated below to further explain the five steps described above. It is applied to the estimation of both summer and winter peak demand reduction associated with various energy efficiency measures.

Total system electrical load or demand in the Electric Reliability Council of Texas (ERCOT) electricity market is used in this example. The ERCOT electricity market is "settled" based on 15-minute intervals. There are 96 intervals in most days. Interval-level data were converted to hourly values to facilitate the estimation and provide a better match of load to hourly temperature data. The top 20 hours of each summer season of each year, *Peak Hour*, were coded 1, and all other hours were coded as 0. Variables representing the hours ending 16:00, 17:00, and 18:00 were included to capture time-of-day factors affecting electricity use. All hours before 2 pm and after 6 pm were assumed to have zero probability of being within the set of peak hours and were eliminated from the dataset to facilitate estimation. Additionally, two variables representing the month-of-year (July and August) were also included. Because summer peak loads are largely determined by air conditioning usage in Texas, a variable was constructed to represent the ratio between the actual temperature in a central location within the ERCOT market (Austin) for a given interval and the highest temperature reading during the given year (*Relative Max Temp*).

The resulting model was thus:

$$\text{Logit}(\text{Peak Hour}) = f(\text{Relative Max Temp, Hour16, Hour17, Hour18, July, August}) \tag{1}$$

This relationship was estimated using R software as a general linear model with a binomial distribution. The estimated coefficients and p-values from the logistic regression are provided in **Table 1**.

---

[1]An application of this probabilistic method in a situation where utility load data can be matched with actual weather and actual electricity prices can be found in [10]. These alternative proposed approaches are adopted from [17].

Table 1. Logistic regression statistical results.

	Estimate	p-Value
Intercept	−39.3331	<0.0001
RelativeMaxTemp	36.1022	<0.0001
Hour16	1.7570	0.000131
Hour17	1.9924	<0.0001
Hour18	1.5439	0.001012
July	0.9284	0.016848
August	1.7722	<0.0001

As we can see from **Table 1**, the coefficient estimates are significant at normally-accepted levels of statistical significance, with the possible exception of the dummy variable denoting the impact of the month of July (relative to the omitted months of June and September).

A unit increase in the relative maximum temperature-the ratio between the actual temperature in a central location within the ERCOT market (Austin) for a given interval and the highest temperature reading during the given year-raises the log of the odds of being included among the peak hours by 36.1022 ceteris paribus.

The coefficient estimate of 1.757 on the variable Hour 16 suggests that the log of the odds of the hour between 3:00 pm and 4:00 pm being among the peak hours (versus the 2:00 pm to 3:00 pm period or hour ending 15:00, the time period not explicitly represented in the model with a variable) is 1.757 time higher, holding all other variables constant. The log odds of setting a peak hour between 4:00 pm and 5:00 pm (Hour 17) versus setting a peak hour between 2:00 pm and 3:00 pm is 1.9924 times higher, holding other variables constant. Similarly, the log odds of the hour from 5:00 pm to 6:00 pm (Hour 18) being among the peak hours is 1.5439 times higher, relative to the omitted period and holding all other variables constant.

For the July and August variables, 0.9284 means the log odds of being a peak hour in July versus being a peak hour in June or September are 0.9284 times higher (which is actually a decrease), and the log odds of being a peak hour in August versus being a peak hour in June are 1.7722 times higher, confirming that summer peaks are most likely to occur in August in Texas.

The coefficient estimates expressed in log odds may be converted to odds ratios, by taking anti-logs.

Once the marginal probabilities are estimated, the probability of each hour of the TMY file being included among the set of peak hours can be calculated. As an example, consider an hour (3:00 pm to 4:00 pm, aka the hour ending 16:00) in August, with an hourly temperature of 100°F, and the annual highest annual temperature being 102°F. The estimated log of the odds ratio of being a peak hour versus being outside the set of peak hours:

$$-39.3331 + 36.1022 \times 100/102 + 1.757 + 1.7722 = -0.4096.$$

Thus the probability of obtaining a peak hour during that time and under those conditions is $\exp(-0.4096)/(1 + \exp(-0.4096)) = 0.4$. This calculation may be performed automatically with R software.

The 20 hours in the TMY file assigned the highest probability of being within the set of peak hours are identified in **Table 2**. Our set of 20 peak hours consists of 7 hours in July and 13 hours in August, all falling within the 3:00 pm to 6:00 pm afternoon time period (*i.e.*, the hours ending 16:00, 17:00, and 18:00). Certainly, temperature prominently determines the probability that an hour falls within the set of 20 peak hours. Yet, the TMY hour ending 17:00 on August 5, 2004 earns the third highest probability, despite having a lower temperature (98.06°F) than some hotter hours (e.g., July 28, 1995 at 16:00 and 18:00). This is because the hour ending 17:00 is more likely to set a peak than the hours ending at 16:00 or 18:00. A probabilistic analysis appropriately takes into consideration both the weather and the time-of-day.

Having estimated the probability of each hour being included among the set of 20 peak hours in this section, we next demonstrate how this information may be used to estimate the impact of energy efficiency measures on peak electricity use when a building energy use simulation model is used to estimate hourly energy consumption using TMY data.

4. Matching the Selected Peak Hours to Energy Efficiency Savings Profiles

To estimate the impact of an energy efficiency measure upon peak demand, we match the hours with the highest

Table 2. Twenty peak hours with the highest probability of being included among the set of peak hours.

Date in TMY File	Hour Ending	Temperature in Degrees F	maxtemp	RelativeMaxTemp	logodds	Probability
7/28/1995	17:00	102.02	102.02	1	−0.3101	0.42309
8/5/2004	16:00	100.04	102.02	0.980592041	−0.40961	0.400743
8/5/2004	17:00	98.06	102.02	0.961184082	−0.86764	0.295746
7/28/1995	16:00	100.94	102.02	0.98941384	−0.92768	0.283395
8/20/2004	16:00	98.06	102.02	0.961184082	−1.10304	0.249171
7/28/1995	18:00	100.94	102.02	0.98941384	−1.14078	0.242177
8/20/2004	17:00	96.98	102.02	0.950597922	−1.24982	0.222731
7/27/1995	17:00	98.96	102.02	0.970005881	−1.39295	0.198937
8/3/2004	16:00	96.98	102.02	0.950597922	−1.48522	0.18464
8/4/2004	16:00	96.98	102.02	0.950597922	−1.48522	0.18464
8/11/2004	16:00	96.98	102.02	0.950597922	−1.48522	0.18464
8/19/2004	16:00	96.98	102.02	0.950597922	−1.48522	0.18464
8/26/2004	16:00	96.98	102.02	0.950597922	−1.48522	0.18464
8/3/2004	17:00	96.08	102.02	0.941776122	−1.56831	0.172457
8/4/2004	17:00	96.08	102.02	0.941776122	−1.56831	0.172457
8/19/2004	17:00	96.08	102.02	0.941776122	−1.56831	0.172457
8/26/2004	17:00	96.08	102.02	0.941776122	−1.56831	0.172457
7/27/1995	16:00	98.96	102.02	0.970005881	−1.62835	0.164056
7/24/1995	17:00	98.06	102.02	0.961184082	−1.71144	0.152977
7/26/1995	17:00	98.06	102.02	0.961184082	−1.71144	0.152977

probability of being among the set of peak hours to those same hours in the output from a building energy use simulation model that used the same TMY data file. The average of the energy efficiency measure's hourly savings over those 20 hours provides an estimate of the savings associated with the efficiency measure coincident with the summer peak.

Application of this approach to a simulation of the savings associated with ceiling insulation and air infiltration in an electrically-heated home in Austin is presented here. We also examine the savings from two lighting-related energy efficiency measures.

A whole-home simulation was developed using Energy Gauge, a simulation software tool that uses a DOE-2 simulation engine [14]. Prototype home characteristics were selected using available data on the construction, occupancy, and equipment characteristics of Texas homes, as listed in **Table 3**. The rows labeled "Ceiling Insulation" and "Air Infiltration" state the base and change conditions.

The simulations assumed differently sized HVAC systems for the analysis of the two weather-sensitive efficiency measures:

• Air infiltration: 2.8 ton air conditioning capacity, 3.5 ton heating capacity
• Ceiling Insulation: 4.3 ton air conditioning capacity, 4.8 ton heating capacity

Table 4 compares estimates of the demand reduction of various scenarios associated with our proposed probabilistic approach with some alternative methods[2]:

• Top 2 Hours of All Peak Months. Select the two hours when the peak hour has most-frequently occurred over the last ten years. Examine impacts during those two hours during every summer weekday during four summer months. Average the impacts over the resulting 170 hours—e.g., the hours ending 17:00 and 18:00 during every summer weekday.

---

[3]These alternative proposed approaches are adopted from [18].

Table 3. Home characteristics inputs used in simulation model.

Input	Value	Source
Conditioned Area	1915 square feet	Weighted average total conditioned square feet of Texas single family detached Single Family Dwelling (SFD) homes.
Site Plan	1 story square, 43'9" × 43'9"	78% of Texas SFD homes are 1 story per 2009 Residential Energy Consumption Survey (RECS) [15]; a square home is agnostic to orientation.
Bedrooms	3	Majority of SFD homes (53%) have 3 bedrooms.
Bathrooms	2	A plurality of SFD homes (41%) have 2 bathrooms.
Foundation	Slab-on-grade, no insulation	Majority (76%) of SFD homes have a slab.
Ceiling Insulation	For Air Infiltration measure R-22. For Ceiling Insulation measure: Base R-2.5, Change R-30.	The average ceiling/wall insulation level for homes existing before 1998 is R-20.51/10.94, per utility baseline studies. It is assumed that all homes built from 1998 on had an average of R-30/13, per International Energy Conservation Code (IECC) 2009 code requirements. Per [15], 78% of Texas SFD homes are pre-1998, and 22% were built on or after 1998. Taking the weighted average U values of insulation, the result is an overall average of U-0.0882/0.0454, or R-11.3/22.0.
Wall Insulation	R-11.3	See above.
Window Area	210 square feet	Per [15], the average Texas home has 14 windows, assuming an average size of 3' × 5' that makes for 210 square feet of windows.
Air Infiltration	For Ceiling Insulation measure: 12.2 ACH50 For Air Infiltration measure: Base 12.2 ACH 50, Change 7.43 ACH50	Based on LBNL's ResDB [16], US average of 0.61 Normalized Leakage (NL) rate for SFDs; per ResDB [16] 0.5NL = 10 ACH50, so 0.61 NL = 12.2 ACH50.
Window U-Value	0.78	Combined the prevalence of single, double, and triple paned glass in Texas SFDs from [15] (58/41/1%) with the average U and solar heat gain coefficient (SHGC) for each pane level from LBNL's RESFEN database [17], excluding windows with high solar gain coatings.
Window SHGC	0.56	See above.
Thermostat Settings	Heating: 71.3°F during the day when someone is home, 67.7°F during the day when no one is home, 69.8°F at night; Cooling: 74.1°F during the day when someone is home, 76.6°F during the day when no one is home, 73.9°F at night.	Weighted average reported thermostat set points from [15]. Times associated with these set points are assumed to be the same as those specified by Energy Star program in US.
Duct Losses	18% total loss	From LBNL's ResDB [16]. National average total duct leakage is 18% of air flow.
Air Conditioning	11.3 SEER	Result of combining the average age of central cooling equipment from [15] with annual shipment-weighted SEER values from the US DOE.
Electric Heater	COP of 1	Fundamental property of electric resistance.

Table 4. Summer peak demand reduction for various efficiency measures from different approaches.

	Ceiling Insulation Austin (kW)	Air Infiltration Austin (kW)	Indoor Lighting Austin (kW)	Outdoor Lighting Austin (kW)
Probabilistic Approach (20 hours)	2.089	0.341	0.062	0
Top 2 Hours of All Peak Months (170 hours)	1.531	0.257	0.087	0
Heat Wave (9 hours)	2.036	0.344	0.056	0
Average Over Peak Period (510 hours)	1.511	0.241	0.069	0

- Heat Wave. The TMY weather files are scanned to locate a three-weekday period that has the highest average temperatures during the peak hours.
- Average Over Peak Period. Estimate a measure's average impact between 1 pm and 7 pm on all summer

weekdays over four summer months. (510 hours)

The second and third alternatives are consistent with the definitions adopted by some state regulatory authorities in the US, as discussed earlier in this paper. The Public Utility Commission of Texas formerly required the *Average over Peak Period* method. Note that two of these three methods ignore the weather information in the TMY file.

The peak demand reduction from two weather-sensitive efficiency measures, ceiling insulation type and air infiltration, is presented in **Table 4**. The estimated average summer demand reduction in Austin for a prototype home using the probabilistic analysis is 2.09 kW for the ceiling insulation efficiency measure and 0.34 kW for the air infiltration efficiency measure.

The demand reduction impacts of two non-weather-sensitive measures, indoor and outdoor lighting in Austin, have also been estimated. For indoor lighting kW savings, we assumed that 30% of the original usage would be saved if energy-saving indoor lighting equipment was installed. Thus an average of 0.062 kW savings could be calculated based on the Energy Gauge home simulation model during 20 summer peak hours. For outdoor lighting, we considered a variety of outdoor lighting equipment and assumed that 5 kW savings when the outdoor light is on is a reasonable deduction. Since none of the summer 20 peak hours occurs at night, the demand reduction associated with the outdoor lighting efficiency measure is 0 kW.

For the two weather-sensitive measures, the definitions involving the highest number of hours yield the smallest estimated peak demand reduction. This is a reasonable result, since including further hours (without regard to the temperature associated with those hours) into a calculation of average impacts shall lower the average and bias the results downward. Our proposed probabilistic method provides estimates which are very similar to the *Heat Wave* method for the summer. The impact of the indoor lighting efficiency measure is greatest under the *Top 2 Hours of All Peak Months Definition*.

Winter peak demand reduction estimates for our proposed approach can be implemented using steps similar to those described above. However, *RelativeMaxTemp* needs to be replaced by *RelativeMinTemp* to represent the ratio between the actual temperature in a central location within the ERCOT market (Austin) for a given interval and the lowest temperature reading during the winter in the year. A *Heat Wave* calculation is not performed for the winter peak. The winter kW savings estimated under three definitions appear in **Table 5**.

The probabilistic approach produces far higher (and more-realistic) winter peak impact estimates for the weather-sensitive efficiency measures. The wide difference in estimates using different approaches can be traced to Texas' climate. Freezing temperatures set the winter peak and are a relatively rare event in this southern state. Deep freezes follow no predictable pattern. That is, one would not expect them to predictably occur during the same month-of-year and hour-of-day year after year. Consequently, the *Top 2 Hours of All Peak Months* performs poorly. Averaging over a prolonged winter peak period (in this case, from 6 am to 10 am and 6 pm to 10 pm) performs very poorly, as well, since many hours with mild temperatures and no need for space conditioning would be introduced into any peak period average.

All 20 of the winter peak hours happen after sunset and before sunrise. Consequently, the demand reduction in the winter for outdoor lighting is 5 kW under two of the three definitions. A lower peak demand reduction estimate is obtained when some daylight hours are included in the definition, as under the *Average Over Peak Period* calculation.

For indoor lighting, similar results are obtained under any of the approaches considered. For the weather-sensitive measures, the probability-based method provides far more-plausible results for a measure's impacts on winter peak for Texas. Extreme temperatures indeed largely coincide with peaks in energy use, so the impacts of an efficiency measure during extreme weather (with adjustments for the time-of-day and month-of-year) should be used when estimating winter peak demand impacts. The use of simple temporal pat-

Table 5. Winter peak demand reduction from different approaches.

	Ceiling Insulation Austin (kW)	Air Infiltration Austin (kW)	Indoor Lighting Austin (kW)	Outdoor Lighting Austin (kW)
Probabilistic Approach (20 hours)	2.253	0.810	0.134	5
Top 2 Hours of All Peak Months (124 Hours)	0.601	0.197	0.183	5
Average over Peak Period (510 Hours)	0.933	0.239	0.105	3.583

terns which ignore temperatures or the averaging over large numbers of hours is inappropriate. It is suspected that in a colder climate where heat waves are a relatively rare event, the naïve application of patterns (without regard for temperature) or averaging to obtain summer peak impacts would similarly lead to implausible results.

5. Conclusions

Utility system planners and energy efficiency program administrators are interested in the impacts of energy efficiency programs at the time of peak demand on a utility system or energy market. Yet, it is not obvious which hour(s) correspond with peak hours when the output from a building energy use simulation model solved with TMY data is examined. Should the hour associated with the highest (or lowest) temperature be used? Should an average of the measure's impacts during the hours and months within which the utility's peak typically falls be used? Should impacts when *design conditions* are experienced be used? Should impacts during consecutive days of extreme weather be averaged? Would a lot of averaging dilute the impact of a weather-sensitive measure?

This paper proposes a formal probabilistic method to address this problem. We select the hours in a TMY weather data file most likely to coincide with a peak hour, based on the temperature, hour-of-day, and month-of-year data contained within the TMY data file and the relationships between these variables and actual load data for a utility system. Logistic regression is used to estimate the relationships based on actual historical data. The estimated relationships and TMY data are used to calculate the probability that an hour represented in the TMY data file would be included among a set of peak hours.

Our proposed approach represents a considerable improvement over existing practices which estimate impacts based solely on extreme temperatures in a TMY file, estimate impacts based upon design-day conditions, averages impacts over a large number of hours within a "peak period", or relies upon typical times of peak occurrence without consideration of the temperature in the TMY file during those hours. When applied to data for Texas, a probability-based approach provides more-realistic estimates of winter peak impacts, relative to two alternatives. When estimating the impacts of an efficiency measure upon summer peak demand, our approach provides impacts similar to the *Heat Wave* approach being used in California.

References

[1]   US Environmental Protection Agency (2006) National Action Plan for Energy Efficiency. US Environmental Protection Agency. http://www.epa.gov/cleanenergy/documents/suca/napee_report.pdf

[2]   Itron, Inc. (2013) DEER Database: 2011 Update Documentation, Appendices.
      http://www.deeresources.com/files/DEER2011/download/2011_DEER_Documentation_Appendices.pdf

[3]   Northeast Energy Efficiency Partnerships (2013) Technical Reference Manual, Version 3.0.
      http://www.neep.org/Assets/uploads/files/emv/emv-products/TRM_March2013Version.pdf

[4]   New York Public Service Commission (2010) New York Standard Approach for Estimating Energy Savings from Energy Efficiency Programs: Residential, Multi-Family, and Commercial/Industrial Measures.
      http://www3.dps.ny.gov/W/PSCWeb.nsf/96f0fec0b45a3c6485257688006a701a/766a83dce56eca35852576da006d79a7/$FILE/TechManualNYRevised10-15-10.pdf

[5]   Public Service Commission of Wisconsin (2010) Focus on Energy Evaluation, Business Programs: Deemed Savings, 2010. http://www.focusonenergy.com/sites/default/files/bpdeemedsavingsmanuav10_evaluationreport.pdf

[6]   State of Illinois (2012) Energy Efficiency Technical Reference Manual.
      http://ilsagfiles.org/SAG_files/Technical_Reference_Manual/Illinois_Statewide_TRM_Version_1.0.pdf

[7]   Xcel Energy (2012) 2012/2013 Demand Side Management Plan, Docket No. 11A-631EG.
      http://www.xcelenergy.com/staticfiles/xe/Marketing/Files/CO-DSM-2012-2013-Biennial-Plan-Rev.pdf

[8]   New Jersey's Clean Energy Program (2007) Protocols to Measure Resource Savings.
      http://www.njcleanenergy.com/files/file/Protocols_REVISED_VERSION_1.pdf

[9]   Efficiency Maine (2013) Residential Technical Reference Manual.
      http://www.efficiencymaine.com/docs/EMT-TRM_Residential_v2014-1.pdf

[10] Zarnikau, J. and Thal, D. (2013) The Response of Large Industrial Energy Consumers to Four Coincident Peak (4CP) Transmission Charges in the Texas (ERCOT) Market. *Utilities Policy*, **26**, 1-6.
      http://dx.doi.org/10.1016/j.jup.2013.04.004

[11] Train, K. (2003) Discrete Choice with Simulation. Cambridge University Press, New York.
      http://dx.doi.org/10.1017/CBO9780511753930

[12] SAS Institute Inc. (1990) SAS/STAT User's Guide, Vol. 1 & 2, Version 6. 4th Edition, SAS Institute Inc., Cary.

[13] Everitt, B. and Hothorn, T. (undated) A Handbook of Statistical Analyses Using R.
http://cran.r-project.org/web/packages/HSAUR/vignettes/Ch_logistic_regression_glm.pdf

[14] Florida Solar Energy Center (undated) Energy Gauge. http://www.energygauge.com/

[15] US Department of Energy (2009) Energy Information Administration, Residential Energy Consumption Survey (RECS). http://www.eia.gov/consumption/residential/

[16] Lawrence Berkeley National Laboratory. Residential Diagnostics Database. http://resdb.lbl.gov/

[17] Lawrence Berkeley National Laboratory. RESFEN. http://windows.lbl.gov/software/resfen/resfen.html

[18] Tetra Tech, Peak Demand Definition Issues. Memorandum to the Public Utility Commission of Texas and Texas Electric Utilities. 2013.

11

Battery Energy Storage System Information Modeling Based on IEC 61850

Nan Wang[1], Wei Liang[2], Yanan Cheng[1], Yunfei Mu[3]

[1]State Grid Tianjin Economic Research Institute, Tianjin, China
[2]State Grid Tianjin Electric Power Research Institute, Tianjin, China
[3]Key Laboratory of Smart Grid of Ministry of Education, Tianjin University, Tianjin, China
Email: 13920692601@163.com

Abstract

This paper discourses the typical ways to access system of the battery energy storage system. To realize the battery energy storage system based on IEC 61850, hierarchical information architecture for battery energy storage system is presented, the general design and implementation methods for device information model are elaborated, and the communication methods of the architecture are proposed. Example of battery energy storage system information model based on IEC 61850 tests that the battery energy storage system information architecture established is feasible.

Keywords

IEC 61850; Battery Energy Storage System; Information Modeling

1. Introduction

Using battery energy storage system in the power system can effectively implement demand-side management, reduce the peak-valley differences between day and night, and strengthen the self-regulating nature of peak-valley load in the regional power grid; Enhance the security and stability of large power grids and power quality levels, improve transmission capacity, and increase reliability; Promote renewable energy assessing to ˌgrids large-scaled. Meanwhile, the battery energy storage system is also an important part of the smart grid. Therefore, energy storage technologies applying in power system will be the development trend of the future power grid [1-3].

Currently, most of the battery energy storage system manufacturers defined device information specifications by their own way, and there were many different types of communication interfaces, which made the battery energy storage system subject to many constraints in information integration, operation control and scheduling management [4]. IEC 61850 uses object-oriented modeling techniques and flexible and scalable communication architecture, has good device characteristics self-describing capability, can meet the requirements of openness and interoperability. IEC 61850 Ed1 aims at communication networks and systems in substations; Ed2 extends coverage to the hydropower, distributed energy and substation automation, etc. which can provide effectively technical means to plug and play device integration, flexible expansion capabilities and interoperability of the information model, and other important functions in the battery energy storage system, reduces the difficulty and cost of system communication and control interfaces standardization work.

2. Typical Battery Energy Storage System Accessing System Mode

According to different functions in power system, battery energy storage system can be classified into different groups, such as the peak and valley battery energy storage system, peak shifting and frequency regulation battery energy storage system, new energy assessing battery energy storage system, and backup power battery energy storage system, etc. [5]. Different functional battery energy storage systems have different mode assessing to power system, which assess by either substations or power plants.

2.1. Constitutions of the Battery Energy Storage System

The battery energy storage system consists of a step-up transformer and a number of storage branches. Each storage branch contains a low voltage breaker and a storage unit. The storage unit is the basic unit of battery energy storage system, which consists of battery pack (BP), battery management system (BMS), and power conversion system (PCS). The constitution of battery energy storage system is shown in **Figure 1**.

2.2. Access to Power System by Substation

The wiring of battery energy storage system assessing to power system by substation is shown in **Figure 2**. The battery energy storage system connects to 10 kV bus in the 110 kV or 35 kV substation through the circuit breaker [6,7]. In the valley period, the battery energy storage system runs on storing energy state, which guarantees

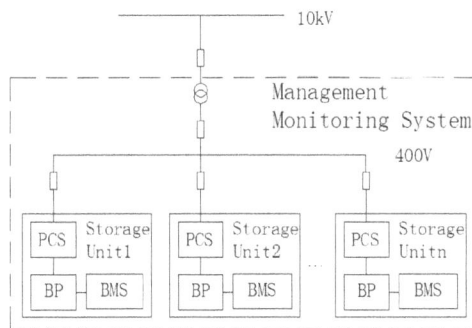

Figure 1. Constitution of the battery energy storage system.

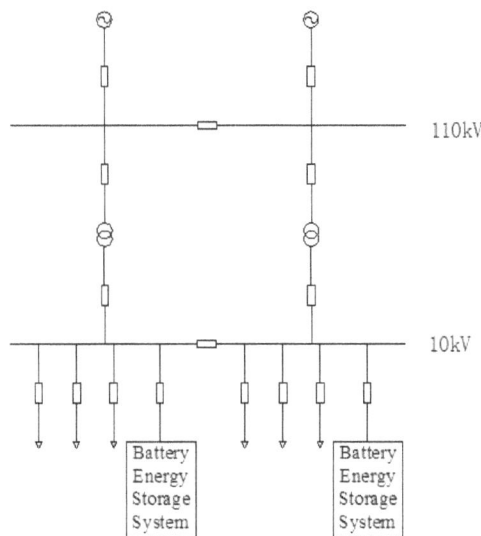

Figure 2. Access to power system by substation.

the load consumption; in the peak period, the battery energy storage system runs on releasing energy state, which guarantees the reliable power supply. Meanwhile, the battery energy storage system can provide power to the important load as backup power when the high voltage side of the substation shutdowns, which improves reliability of power grid.

2.3. Access to Power System by Power Plant

Since the gradual depletion of fossil fuels and the greenhouse effect caused by the growing phenomenon of global warming, renewable energy has attracted great attention. However, intermittent renewable energy, such as wind energy and solar energy, is not continuous and stable. The battery energy storage system is used with renewable energy generation systems to improve the imbalance of the renewable energy power generation system and electric power load, the quality of power supply, and the renewable energy power system stability. The wiring of battery energy storage system assessing to power system by power plant is shown in **Figure 3**.

3. Information Structure Design of the Battery Energy Storage System

The information structure of the battery energy storage system can be classified into three layers, including the device layer, the bay layer and the management layer. The information structure of the battery energy storage system is shown in **Figure 4**.

3.1. The Device Layer

The device layer is the basic of the battery energy storage system; it is composed of physical equipment, including battery pack, DC breaker, DC current convertor, measuring and monitoring equipment, AC breaker and step-up transformer, etc. The measuring and monitoring equipment, such as current transformer, potential transformer, is in charge of battery energy storage system measuring and operating data monitoring, which provides data support for real-time analysis of battery energy storage system. The AC breaker and step-up transformer connect the battery energy storage system with the external power grid.

3.2. The Bay Layer

The bay layer is the bridge of information exchange between device layer and management layer. On the one

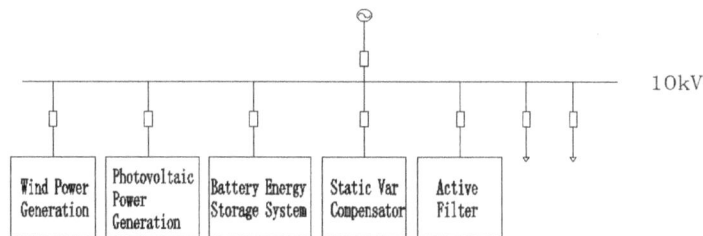

Figure 3. Access to power system by power plant.

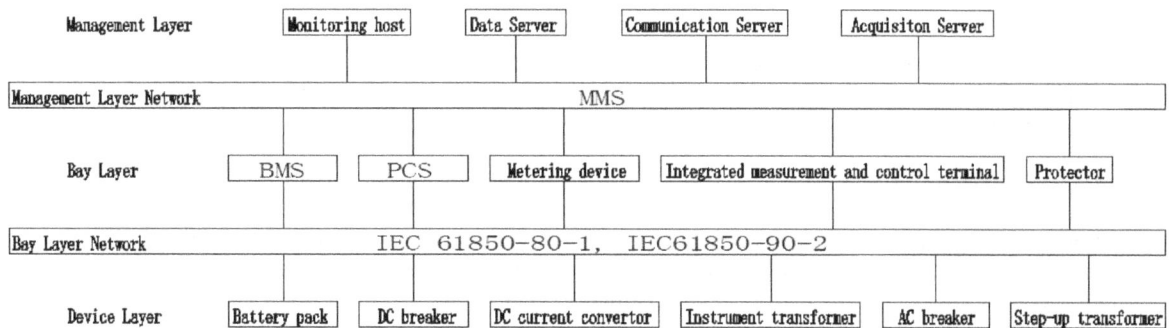

Figure 4. Information structure of the battery energy storage system.

hand, the bay layer receives measurement data and operating status information collected by the device layer, preprocesses the data and exchanges them with the management layer; on the other hand, it receives, analyzes and delivers the control information of the management layer, realizes the management of the device. The devices in the bay layer include battery management system (BMS), power conversion system (PCS), integrated measurement and control terminal and protectors. They typically have the extended human-computer interaction interface, which are convenient for parameters configuration and operation debugging. The BMS mainly completes the function of battery monitoring, running alarm, self-diagnosis and parameter management, is able to display the battery running state, and realizes the function of storage and query of historical data. The PCS realizes battery energy storage system connecting to the grid or islanding operation according to the orders that sent by either the management layer or the bay layer. Integrated measurement and control terminal realizes the function of data collecting and control orders sending. The protectors realize the function of fault protection and status alarm of the battery energy storage system.

3.3. The Management Layer

The management layer is consisted of the monitoring host, operator station and telemechanism communication device, provides the man-machine interface, realizes the function of managing and controlling the device layer and the bay layer device, formats the substation monitoring& management center, and communicates with the remote monitoring & controlling center.

4. The Design of Battery Energy Storage System Information Model and Communication Methods

The battery energy storage system uses the information model based on IEC 61850 in order to realize large capacity data monitoring. Basic model structures and communication methods of the power system device follow the IEC 61850 standard Part 5, various IED device configuration descriptions follow the IEC61850 standard Part 6, the information model of automatic control, monitoring, protection and other functions follow IEC 61850 standard Part 7-4. The IEC 61850 standard (Ed.2) Part 7-420 defines the information model of distributed energy. The information model of the battery energy storage system follows the above standards. The devices that haven't meet IEC 61850 standard can be equipped with the appropriate IED to complete rapid configuration and integration of the information model [8-10].

The packets exchanged between the device layer and the bay layer including trip command, alarm signal, meter and operational data collected by sensors. The device layer communicates with the bay layer through IEDs. The BMS monitors the battery cell voltage, temperature and alarm. The PCS uses monitoring data and system control strategies to achieve closed-loop controlling of converter. Information exchange follows IEC 61850 standard Part 80-1 & 90-2, uses fiber Ethernet to transmit data based on TCP/IP protocol in order to ensure that the communication network has high reliability between the device layer and the bay layer.

The packets exchanged between the bay layer and the management layer include automatic control data, protector value, time synchronization messages and control command packets, which have high requirements for information transmission security and real-time communication, can communicate by fiber optic network.

5. Examples of Battery Energy Storage System Information Modeling Based on IEC 61850

The information model of AC breaker and step-up transformer can use the model designed in ordinary substation. The information model of the battery energy storage system based on IEC 61850 is shown in **Figure 5**. Main logical nodes are listed in **Table 1**.

6. Summary

The openness and interoperability of IEC 61850 makes communication and control interface standardization of battery energy storage system feasible. This paper discourses the typical accessing ways of the battery energy storage system. To realize the battery energy storage system based on IEC 61850, hierarchical information architecture for battery energy storage system is presented, the general design and implementation methods for device information model are elaborated, and the communication methods of the architecture are proposed.

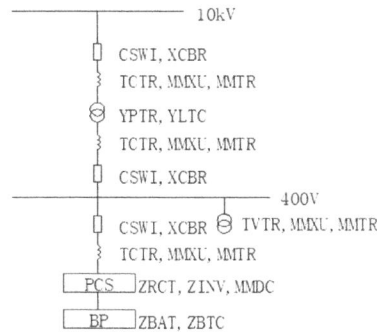

Figure 5. Information model of the battery energy storage system based on IEC 61850.

Table 1. Logical nodes and corresponding function.

Logical node	corresponding function	Explanation
ZBAT	Battery systems	Remote monitoring and control battery systems
ZBTC	Battery charger	Remote monitoring and control battery chargers
ZRCT	Rectifier	Define the characteristics of rectifier
ZINV	Inverter	Define the characteristics of inverter
MMDC	DC measurement	DC voltage, current, power, and resistance
MMXU	AC measurement	AC voltage, current, active and reactive power
CSWI	AC breaker controller	Describe the controller for operation of AC breaker
XCBR	AC breaker	Define the characteristics of AC breaker
MMTR	Meter	Meter information
YPTR	Transformer	Define the characteristics of transformer
YLTC	Tap changer	Define the characteristics of tap changer
TVTR	Voltage transformer	Define the characteristics of voltage transformer
TCTR	Current transformer	Define the characteristics of current transformer

Example of battery energy storage system information modeling based on IEC 61850 tests that the battery energy storage system information architecture established is feasible.

References

[1]   Zhang, W.L., Qiu, M. and Lai, X.K. (2008) Application of Energy Storage Technologies in Power Grids. *Power System Technology*, **32**, 1-9.

[2]   (2011) Electrical Energy Storage. IEC, Geneva.

[3]   Rastler, D. (2010) Electricity Energy Storage Technology Options: A White Paper Primer on Applications, Costs, and Benefits. Electric Power Research Institute (EPRI), Palo Alto.

[4]   Deng, W., Pei, W. and Qi, Z.P. (2013) Microgrid Information Exchange Based on IEC 61850. *Automation of Electric Power Systems*, **37**, 6-11.

[5]   Zhou, L., Huang, Y. and Guo, K. (2011) A Survey of Energy Storage Technology for Micro Grid. *Power System Protection and Control*, **39**,147-152.

[6]   Lu, Z.G., Wang, K. and Liu, Y. (2013) Research and Application of Megawatt Scale Lithium ion Battery Energy Storage Station and Key Technology. *Automation of Electric Power Systems*, **37**, 65-69, 127.

[7]   Jin, Y.D., Song, Q., Chen, J.H., *et al.* (2010) Power Conversion System of Large Scaled Battery Energy Storage. *Elec-*

tric Power, **43**, 16-20.

[8]   Cao, J.W., Wan, Y.X. and Tu, G.Y. (2013) Information System Architecture for Smart Grids. *Chinese Journal of Computers*, **36**, 143-167.

[9]   Huang, X.M. (2012) Information Modeling and Distributed Execution Mechanism for Flywheel Energy Storage System in Micro Grid. *Power System and Clean Energy*, **28**, 54-63.

[10]   Wang, D.W., Di, J. and Zhang, C.M. (2012) Information Modeling and Implementation for Status Monitoring IED in Substation Based on IEC 61850. *Automation of Electric Power Systems*, **36**, 81-86.

12

LVDC: An Efficient Energy Solution for On-Grid Photovoltaic Applications

Anis Ammous[1,2], Hervé Morel[3]

[1]Engineering School of Sfax (ENIS), University of Sfax, Sfax, Tunisia
[2]College of Engineering and Islamic Architecture, Umm Al Qura University, Mecca, Saudi Arabia
[3]Université de Lyon, INSA Lyon, Lab. AMPERE, CNRS, Villeurbanne, France
Email: anis.ammous@enis.rnu.tn

Abstract

In this paper some photovoltaic, PV, conversion chains architectures for on-grid applications have been proposed and the advantage of the direct use of a Low Voltage Direct Current (LVDC) bus for the DC loads has been shown. The evaluation of the efficiency of the proposed chains compared to the classical one was performed. It is shown that LVDC use instead of standard AC plugs, in numerous applications, is promising in future. The registered annual saved energy can exceed 25% of the PV generated energy. This important rate, the need of better services at lower economic cost and environmental burden will incite to make reflection about industry and supplies' future standards.

Keywords

Solar Energy, PV Panel, On-Grid PV Systems, LVDC, Energy-Efficiency

1. Introduction

Nowadays, world uses a high amount of energy. Moreover, electricity is becoming an essential part of the society, perhaps because electricity may be seen as a green energy. Concerning the energy production, a new requirement concerns a low rate of emissions of green-house gases and particularly the carbon dioxide (CO_2) into the atmosphere to limit the global warming.

So, in recent years an important increase of renewable energy sources like the solar photovoltaic energy has been seen. The use of these energies locally produced in offices and houses will continue to increase in future and no one can predict its amplitude.

The solar PV systems connected with public power systems (On-Grid Solar Energy Systems) are the most current way to use renewable energy. This solution avoids the use of accumulators for the energy storage. A PV cell directly converts some solar energy in to some direct-current (DC) electricity but additional converters are required to connect them to the grid. The efficiency of the global PV system depends on PV cells' efficiency, converters' efficiencies and global conversion chains' architecture [1].

Classically, in housing and office applications, the use of the electric energy is achieved from the alternative current plugs even if PV panel systems are locally connected to the power grid. This process can considerably increase system losses especially when DC electricity is used at the load levels.

In practice, electronic appliances, such as computers, gaming consoles, printers, economic or LED lights, televisions and so on need DC supplies. Additional AC to DC converters are needed in such equipments [2]-[5].

The main object of this paper is to show the importance of the saved energy values when LVDCs are used. As it is described in following section of the paper, the issues are important. A LVDC bus in a PV system is regulated by acting on converter control.

In the first section of the paper the state-of-the-art DC current supplies are described and the need of power factor correction, PFC, systems to improve absorbed current waveforms from the grid are recalled.

In the second section, we have proposed practical PV conversion chains which provide a direct use of the DC current for DC loads applications. The PV system is connected to the power grid and no accumulator battery is used. The different converters modeling are performed using non ideal averaged modeling techniques based on the switching cell configuration [6] [7]. This technique was considered to be efficient especially for semiconductor losses estimation as it is proved by refined simulations and experiments [7]. The averaged modeling of the different converters constitutes an important way to efficiently model a complex PV system. The refined simulation of the global conversion chain, in a circuit simulator, is not easy and tedious because many semiconductor devices operating with high switching frequency are used.

The last section treats the quantitative evaluation of the efficiency of the different proposed PV chains compared to the classical one. A practical profile of consumed power evolution in an office with a PV panel (600 W peak) generating power magnitude during one day is considered. An important saved energy when a DC voltage bus is used directly for DC loads applications has been registered. The practical application was studied for two different geographic places, with two different solar profiles, Lyon in France (a standard solar radiation location) and Sfax in Tunisia (a high solar radiation location). Annual saved energy can exceed 25% of the annual PV generated energy. It is noticeable that the importance of the performed studies is even more interesting when the main consumed power is achieved during daytime.

2. Direct Current Supplies

2.1. Power Factor Correction Systems

Most of loads daily used in offices and at home require direct current power supplies to operate. Despite this, these loads use the standard alternative current grid as a first energetic input. Therefore, this operation needs alternative to direct current conversion (Switch Mode Power Supplies SMPS as shown in **Figure 1**) which is not without consequences on the supply network as well as the system sizing and the global losses [8]-[10].

One problem with switch mode power supplies (SMPS) is that they do not use any form of power factor correction. Moreover, the input capacitor C_{IN} only loads when V1 is greater than the voltage across the capacitance V_{CIN}. If C_{IN} is designed from the input voltage frequency, the current is looked much closer to the input waveform (load dependent). However, any little interruption on the mainline will cause the entire system to react negatively. In saying that, in the designing of a SMPS, the hold-up time for C_{IN} is designed to be greater than the period of V_{IN}, so that if there is a glitch in V_{IN} and a few cycles are missed, C_{IN} will have stored energy enough to continue to power its load.

The power factor, PF, of any component consuming an instantaneous power $p(t)$ in a cyclic operation of period T may be defined as

$$PF = \frac{P}{S} \tag{1}$$

where P, is the real power (W) given by,

$$P = \langle p \rangle = \frac{1}{T}\int_0^T p(t)\,\mathrm{d}t \tag{2}$$

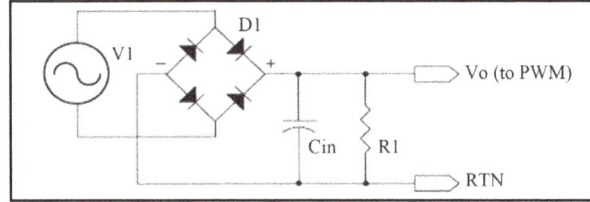

Figure 1. SMPS Input Without any PFC.

and S, is the apparent power (VA) is given by,

$$S = \sqrt{\frac{1}{T}\int_0^T i(t)^2 \, \mathrm{d}t}\sqrt{\frac{1}{T}\int_0^T v(t)^2 \, \mathrm{d}t} \tag{3}$$

where $i(t)$ is the instantaneous current flowing through the component and $v(i)$ its instantaneous across voltage.

The power Factor PF is expressed as decimal number between zero and one.

The typical switched mode power supply should have a power factor of about 0.6, therefore having considerable odd-order harmonic distortion. That reduces the real power available to operate the device. To operate a device with these inefficiencies, the power company must supply additional power to make up for the loss.

This is why a Power Factor Correction (PFC) on the device side has become an important part of the final power system design for so many products.

There is a lot of standards (For example, EN 61000-3-2 in Europe) to drive power consumption near a power factor of 1 and keep the total harmonic distortion to a minimum. Depending on the output power and the designer's needs, a SMPS can be designed with either a discontinuous or continuous mode standalone PFC controller, or a continuous PFC/PWM mode device (**Figure 2**).

Many types of PFC are used in practice, the best known one is the boost converter. The boost converter topology is used to accomplish this active power-factor correction in many discontinuous/continuous modes (**Figure 3**). The boost converter is used because it is easy to implement and it operates well.

The input of the converter is a full-rectified AC line voltage. No bulk filtering is applied following the bridge rectifier. So the input voltage of the boost converter varies (at twice the line frequency) from zero to the peak value of the AC input.

2.2. Classical Photovoltaic Systems Connected to the Grid

The photovoltaic electricity has initially been developed for standalone applications without connection to a power grid. For example, it was the case of satellites or isolated housings. Nowadays, it is found in various power applications such as personal calculators, watches and other objects of daily use. Indeed, the electricity produced by photovoltaic cells can supply various individual continuous loads without difficulty. More recently, with the emergence of photovoltaic systems connected to the grid, the PV has significantly expanded as a way to generate electricity.

Connecting PV systems to the grid (**Figure 4**) has many advantages compared to standalone installations, for example:

- The produced current is conforming to the current network (230 V, 50 Hz, in Europe) and can be used directly by consumers (appliances, light bulbs, etc.).
- Solar power users do not need a separated network;
- In case of overproduction (or inversely, over consumption) balancing energy is performed automatically through the network connection;
- A "virtual storage" of energy is provided by the electric network and batteries are superfluous;

Figure 4 shows one of classical configurations used to connect PV panels to the standard low voltage grid without the need of transformer. The DC/DC converter, Converter #1, allows extracting the maximum power from the PV panels when climatic conditions like the solar radiation changes. The Maximum Power Point Tracking (MPPT) algorithm is often used by acting on the DC/DC converter control.

The DC/AC converter, Converter #2, can transfer the PV generated power to the grid and insures the regulation of the input DC voltage value (400 V) of the inverter. This value allows obtaining easily the standard 230 V

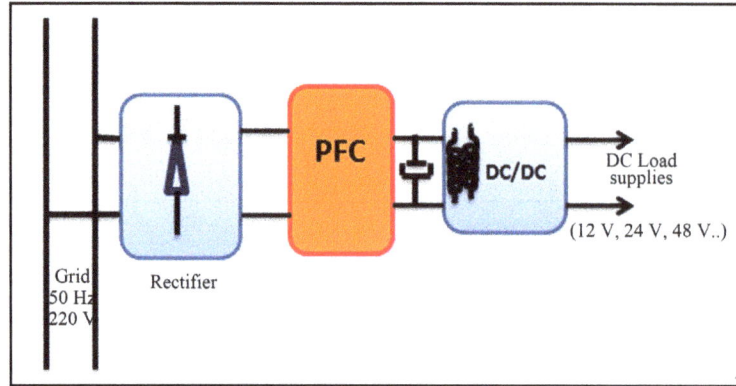

Figure 2. An SMPS with a power factor correction bloc.

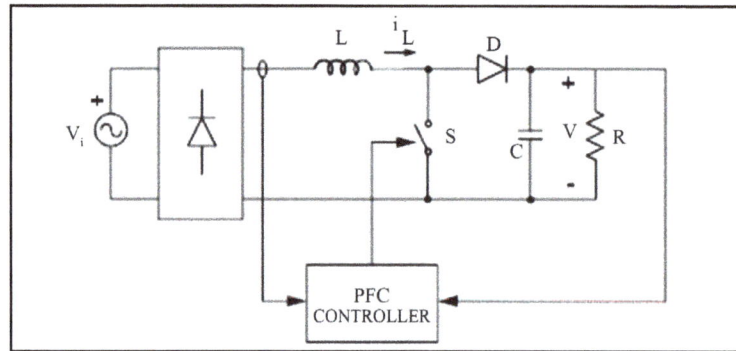

Figure 3. PFC circuit based on the boost converter.

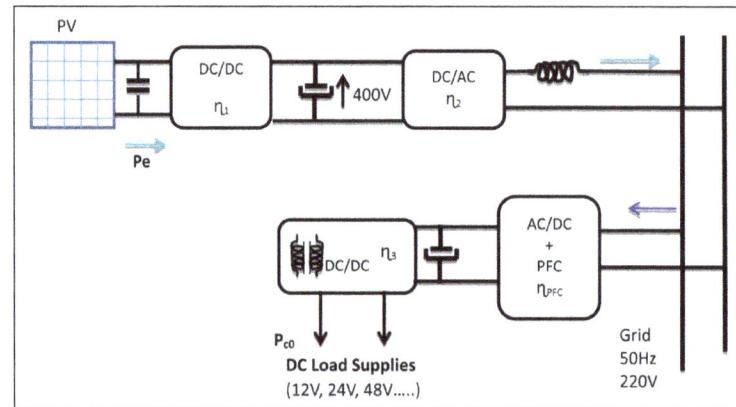

Figure 4. Classical configuration of the PV panel and DC load connected to the grid.

AC. The injected current to the grid have a quasi-sinusoidal waveform and is in phase with the imposed grid voltage. The power factor is near 1.

As previously described, a power factor correction stage below the rectifier is frequently used in power DC supplies. The output stage, the DC/DC converter, converter #3, often with a galvanic isolation (using HF transformer) is used to adjust the PFC output voltage to the direct current loads.

Table 1 resumes the used converters, in the different studied PV chains, and their specifications.

The classical configuration in **Figure 4**, shows that a portion of photovoltaic generated power should pass through four converters to honor the load demand. So, the system efficiency is low in the case of DC loads and a significant part of the PV power (Pe) is lost.

Table 1. Used converters and their specifications.

Converter name	Nature	Efficiency symbols and typical values	Reversibility
Converter #1	DC/DC, Buck or Boost, (Boost in our case)	$\eta_1 = 0.95$	Irreversible
Converter #2	DC/AC	η_{20} when PWM inverter η_{2r} when PWM rectifier $\eta_{20} = \eta_{2r} = 0.9$	Reversible
Converter #3	DC/DC Buck, often with HF transformer	$\eta_3 = 0.94$	Irreversible
Converter #4	DC/DC, Buck in system #3	η_{4bk} when Buck η_{4bs} when Boost	Irreversible
	Buck-boost in system #2	$\eta_{4bk} = \eta_{4bs} = 0.92$	Reversible
Converter #5	AC/DC + PFC	$\eta_{pfc} = 0.89$	Irreversible

3. On-grid Photovoltaic System Configurations and the Direct Current Use

3.1. Proposed On-Grid PV Chains

In this section, some photovoltaic architectures with a direct current chain form PV to load. The object of this study is to show that using direct current from a DC bus (for DC loads), without need of any alternative current grid, is very economic and increase the global system efficiency.

The first proposed architecture is shown in **Figure 5**. The idea is to use directly the regulated DC bus (400 V). The DC/DC (η_3) converter adapts LVDC loads to this bus.

The second architecture is shown in **Figure 6**. In this structure, a second regulated bus (<120 V) is created which is in agreement with actual standards. Indeed, 120 V is considered to be the secure limit for people in a dry environment [11]. For wet environment this limit is 60 V. The safe low voltage DC bus requires the use of an additional reversible DC/DC converter (η_4). When generated PV power at the LVDC bus is greater to the consumed power, the additional converter operates as a boost. Otherwise it operates as a buck converter.

It is obvious that the second solution decrease global system efficiency but has the advantage to generate a safe low voltage bus allowing more security for people.

Figure 7 shows the third proposed PV architecture. It consist to create a Safe Low Voltage DC bus (<120 V) as the LVDC bus. The additional DC/DC converter (η_4) is a buck converter.

As well as the second solution, the proposed architecture sweet well with standards in office and domestic applications because its ability to be easily connected to the grid.

3.2. Analysis of the Power Balance in the Proposed PV Architectures

The object of this section is to compare the proposed PV architecture to the classical one which is widely used today. η_{20} and η_{2r} are the DC/AC converter efficiencies when it operates in PWM inverter mode and PWM rectifier modes respectively.

η_{4bk} and η_{4bs} are the DC/DC converter efficiencies when it operates in buck mode and boost mode respectively.

P_e is the maximum output power of the PV panels. The efficiency of the MPPT algorithm is included in the DC/DC efficiency value (η_1).

P_{c0} is the DC load consumed power. P_i is the injected power to the grid after honoring the loads. This power is positive if the PV generated power is greater than consumed one (obviously, different converter efficiencies are taken into account). Otherwise P_i is negative.

Table 2 resumes the expressions of the injected power P_i. These expressions are function of the input power P_e, the consumed power P_{c0} and the different converter efficiencies. The classical system is used as a reference structure.

To compare the proposed PV architectures to the classical one and to evaluate the magnitude of the lost in the different converters it is necessary to develop accurate models of these converters. The proposed models should take into account the static and dynamic losses in the different power semiconductor devices.

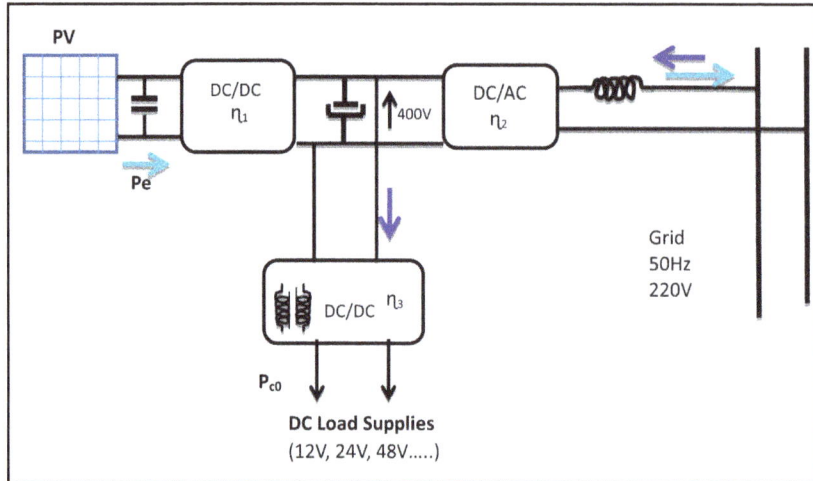

Figure 5. First proposed PV system architecture for DC supplies (**System #1**).

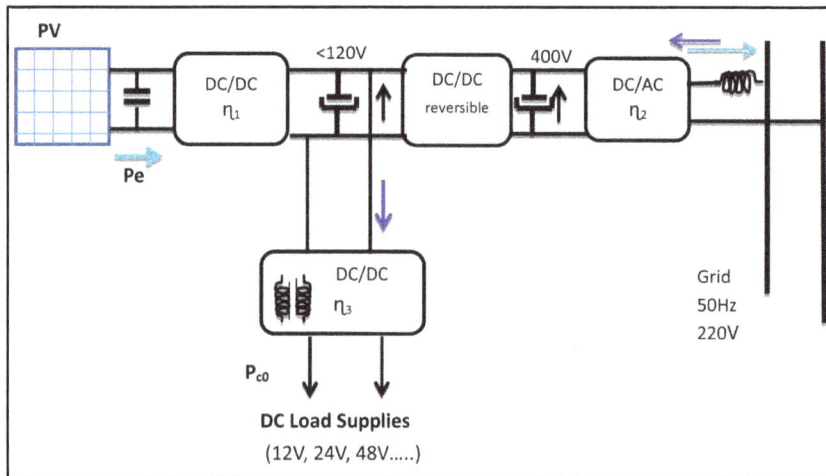

Figure 6. Second proposed PV system architecture for DC supplies (**System #2**).

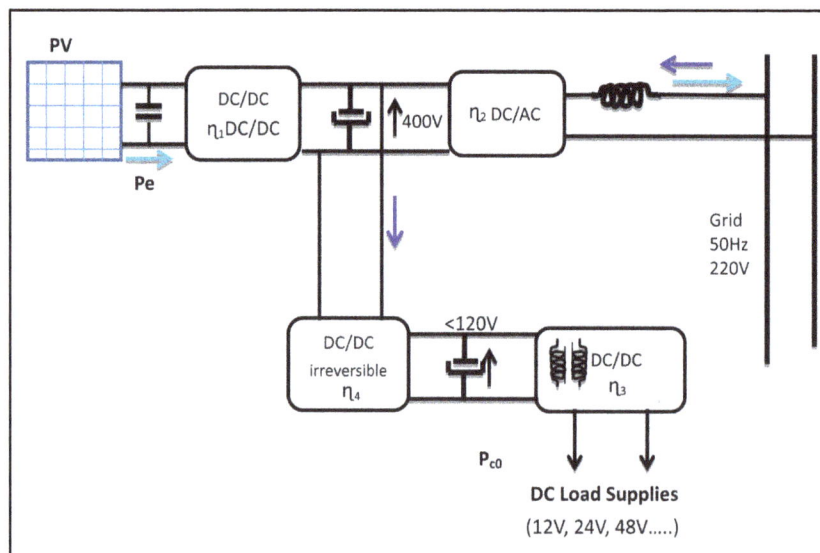

Figure 7. Third proposed PV system architecture for DC supplies (**System # 3**).

Table 2. Expressions of the injected power P_i to the grid.

Conversion chain	P_i (≥ 0)	P_i (<0)
Classical system	$\eta_1\eta_{20}P_e - (1/\eta_3\eta_{PFC})P_{C0}$	$\eta_1\eta_{20}P_e - (1/\eta_3\eta_{PFC})P_{C0}$
System #1	(If $P_e\eta_1/\eta_3 \geq P_{C0}$) $\eta_1\eta_{20}P_e - (\eta_{20}/\eta_3)P_{C0}$	(If $P_e\eta_1/\eta_3 < P_{C0}$) $(\eta_1/\eta_{2r})P_e - (1/\eta_{2r}\eta_3)P_{C0}$
System #2	(If $P_e\eta_1/\eta_3 \geq P_{C0}$) $(\eta_1\eta_{20}\eta_{4bs})P_e - (\eta_1\eta_{4bs}/\eta_3)P_{C0}$	(If $P_e\eta_1/\eta_3 < P_{C0}$) $(\eta_1/\eta_{2r}\eta_{4bk})P_e - (1/\eta_{2r}\eta_{4bk}\eta_3)P_{C0}$
System #3	(If $P_e\eta_1/(\eta_3\eta_{4bk}) \geq P_{C0}$) $(\eta_1/\eta_{20})P_e - (1/\eta_{20}\eta_{4bk}\eta_3)P_{C0}$	(If $P_e\eta_1/(\eta_3\eta_{4bk}) < P_{C0}$) $(\eta_1/\eta_{2r})P_e - (1/\eta_{2r}\eta_{4bk}\eta_3)P_{C0}$

4. LVDC Advantage in On-Grid PV Generation Systems

4.1. Application Description

To show the efficiency of the proposed PV systems compared to the reference structure, a 600 W PV panel is chosen. In our case this panel is destined to office applications where the PV power is generated locally. It can be used to supply administrations, laboratories and domestic applications too.

The next study was achieved for two cities, the first one is Lyon in France (Europe) and the second is Sfax in Tunisia (North Africa).

In Lyon the average annual sunshine duration is 2006 h and the average annual radiant energy is about 1400 kWh/m²/year.

In Sfax the average annual sunshine duration is 3000 h and the average annual radiant energy is about 2400 kWh/m²/year [12].

Assuming a PV panel efficiency equals to 0.15, the annual electrical energy registered in Lyon is about 210 kWh/m²/year and the electrical energy obtained in Sfax is 360 kWh/m²/year.

Figure 8 shows the considered power profile of the office load consumption and the PV locally generated power during a good weather day. Indeed, around 13 o'clock, the maximum PV power is generated and a considerable decrease on office consumption was registered because of office staff breaks.

The injected power into the grid (P_i) is calculated and the daily evolution of this power is registered.

Figure 9 shows the evolution of P_i obtained by the different PV systems (the classical and the proposed chains conversion). At this stage of the study, constant converter efficiencies are considered (**Table 1**). These efficiencies values are considered as an example of typical values for the chosen converters operating at nominal power.

In order to evaluate the efficiency of each studied, chain conversion, architecture, the difference ΔP is integrated during one day. ΔP is the difference between the injected power obtained by each proposed PV system and the injected power obtained by the classical system.

So,

$$\Delta P(k) = P_i \text{ generated by the proposed system } (k) - P_i \text{ generated by the classical system} \tag{4}$$

where $k = 1, 2, 3$ corresponds to the index of the proposed chain systems. The energy balance between the PV system (k) and the reference PV system, is defined as,

$$W(k) = \int_{day} \Delta P(k) \tag{5}$$

When $W(k)$ is positive, the use of PV system (k) is better than the classical system. In the later case, $W(k)$ is the saved energy. Contrarily, if this energy is negative, the reference PV system is the more efficient architecture solution.

With the converter efficiency values considered in **Table 1** and the injected power Pi expressions (**Table 2**), the different PV converter energy balances $W(k)$ are given in **Table 3**. These values are obtained in the case of a given weather and fixed converters efficiencies value.

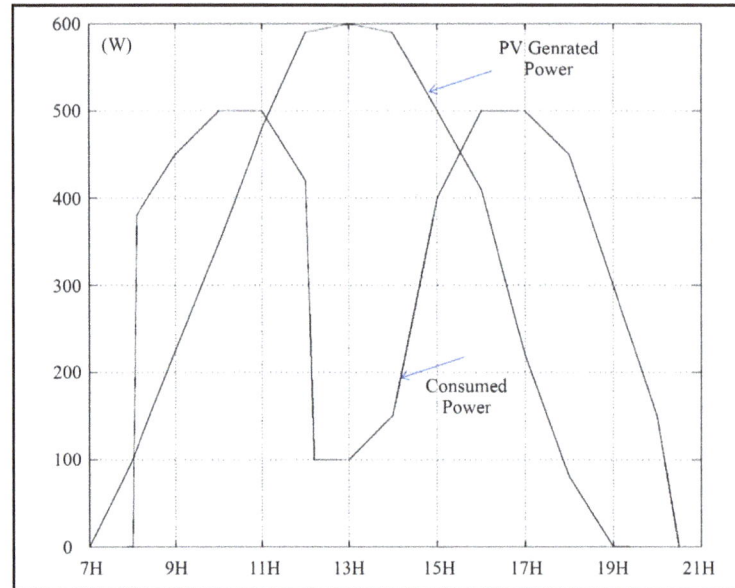

Figure 8. PV generated power and load consumed power during a good weather day.

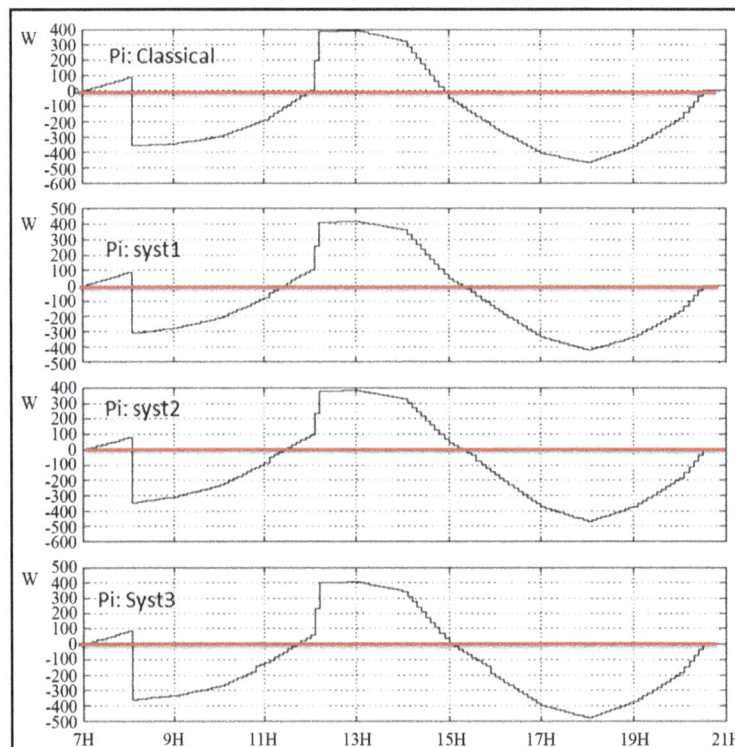

Figure 9. The injected power (Pi) evolutions for the different PV system configurations.

All the energy values given in **Table 3**, are positive, so the proposed converter chains architectures are always better than the reference PV system configuration. For a given PV system architecture, the energy value depends mainly on radiant energy and converters efficiencies.

For further accuracies of the efficiency analyses, we have proposed to take into account the non-ideal behavior of the different converters in the proposed PV converter chains.

Table 3. Energy balance $W(k) = \int_{day} (P_i \text{ system\#}k - P_i \text{ classical system})$ of the proposed PV systems.

Proposed PV chain	Energy balance $W(k)$ Wh/day
System #1	750
System #2	450
System #3	240

4.2. Proposed Converters Model

To estimate the energy efficiency of the different studied circuit. The classical approach consists in using a circuit simulator and a model of the complete circuit included the power switches, the filtering elements and also a layout parasitic representation that dramatically act on the converter losses. However such an approach demands lot of efforts and a high simulation cost. Fortunately there exist other simplified approaches based on averaged models that enable to reduce the simulation cost. The proposed analyses are based on the representation of the converters in the different photovoltaic chains by non-ideal averaged models. The model takes into accounts the semiconductor non linearity. Static and switching characteristics of the different devices are considered [6] [7]. The DC bus inductance Ls, the diode inverse recovery phenomenon and the dead time of the governed signals are taken into account too. L is the storage inductance in the converter (**Figure 10**).

The proposed averaged model corresponding to the PWM-cell of **Figure 10** is shown in **Figure 11**. The controlled voltage sources (V1) and current source (I1) are given by. $V_1 = \langle U_{as} \rangle$ and $I_1 = \langle I_{e2} \rangle$.

Where $\langle U_{as} \rangle$ and $\langle I_{e2} \rangle$ are the time averaged values of the instantaneous terminal waveforms $U_{as}(t)$ and $I_{e2}(t)$ respectively over the switching period T_s.

The accuracy of the proposed averaged models was largely proved by comparisons to refined simulations and experimental tests [6] [7].

The non-ideal averaged model estimates the dissipated power in the different semiconductor devices. This is the first important criterion to use this type of models in an application. The second criterion is the improvement in simulation cost (a ratio of about 10^4) in favor of the non-ideal averaged model with respect to refined models.

The averaged models of the different proposed PV systems are implemented in Matlab simulator. They can be implemented in any circuit oriented simulation tools too.

4.3. Energy Balance Study

The efficiency of each converter depends on different considerations like the semiconductor types and the input power magnitude.

Considering the following semiconductor devices used in the different proposed converters:

Stth2002 diode and stw52nk25z MOSFET are for the very low DC voltage bus. Stta2006 diode and the irfps43n 50 k MOSFET are for the low DC voltage bus. In this application, the active devices are driven by a 40 kHz switching frequency value. The DC/DC converter (η_3) is assumed to have the same architecture in the different PV chains. η_{PFC} is the efficiency of a diode rectifier converter followed by a boost converter.

Figure 12 shows the evolution of the different converter efficiencies as a function of transferred power. When the delivered power is very low with regards to converter nominal power, the converter efficiency considerably decreases.

In the same power conditions as defined in the last section, the integral of $\Delta P(k)$ quantity during a good weather day (**Figure 13**) is registered.

Taking into account the converter nonlinearities it is noticed that saved energy $W(k)$ values are different from estimated values when converters efficiency are considered as constants.

The best saved magnitude value is registered by system number 1. This is predictable, because of the use of a minimal number of converters and semiconductors. The disadvantage of this architecture is the value of the DC bus voltage (400 V) which not satisfy security standard for people. This doesn't exclude the use of this kind of DC bus in future but with additional security devices.

In the case of a very bad weather day (**Figure 14**), the energy balance $W(k)$ value can be negative and classic-

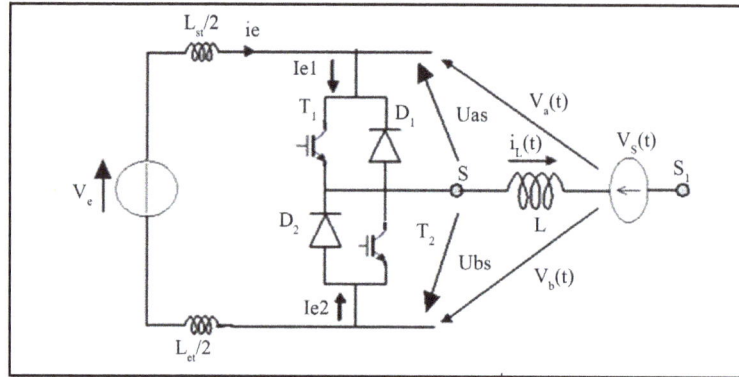

Figure 10. PWM-cell.

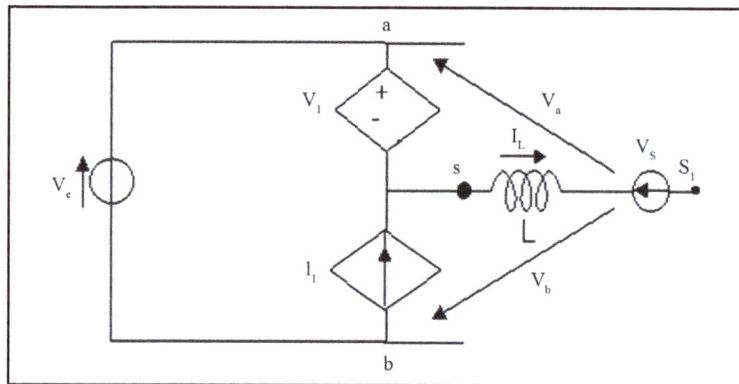

Figure 11. Proposed averaged model of the PWM-cell.

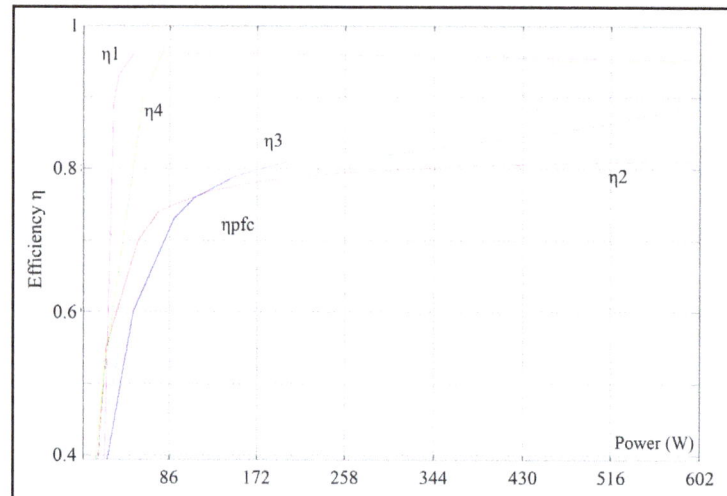

Figure 12. Converter efficiencies as function of power magnitude.

al configuration gives the best global PV system efficiency as shown in **Figure 15**.

It is clear that the efficiency of the proposed PV converter chains depends mainly on weather conditions. In the following section, the global energy balance of the different architectures during one year in the two considered cities, Lyon and Sfax is considered.

Four different average weather states, in the year, have been considered as shown in **Figure 16**. The annual produced PV panel energy is about 1.4 MWh in Lyon. The four averaged energy values are 5500 Wh/day, 4500 Wh/day, 3000 Wh/day and 1100 Wh/day.

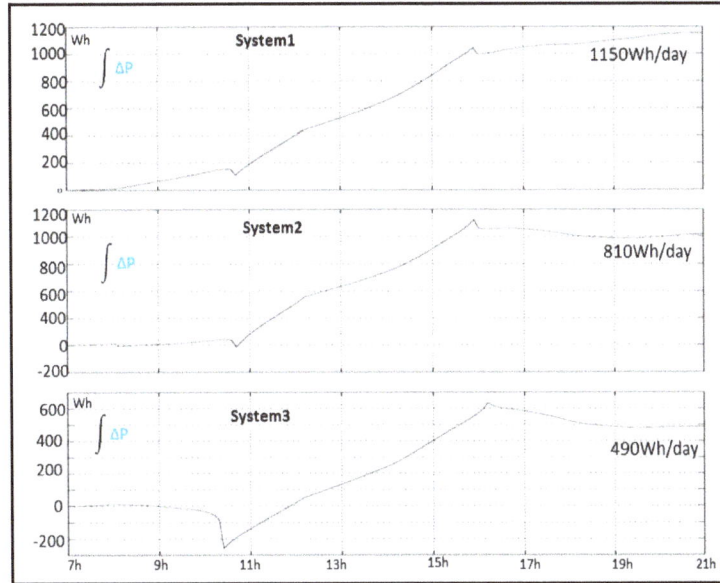

Figure 13. $W(k)$ energies evolution during a good weather day.

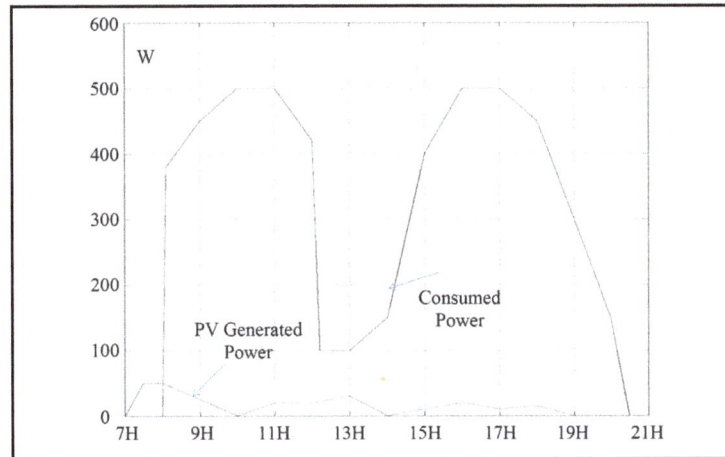

Figure 14. PV generated power and Load consumed power during a very bad weather day.

The daily averaged energy balance of the different proposed converter chains $W(k)$ is illustrated in **Figure 16** too. The annual saved energies are equals to 404 kWh, 278 kWh and 176 kWh when using the three defined systems respectively. The saved energy obtained by the proposed system number 1 for example, corresponds to 28.8% of the total PV annual produced energy. This rate is very important and should be considered for future PV converter chain topologies.

In the case of Sfax city the four assumed averaged PV produced energy (7500 Wh/day, 6500 Wh/day, 5500 Wh/day and 4000 Wh/day) are shown in **Figure 17**. The annual produced PV panel energy is about 2.3 MWh.

The annual saved energies are equals to 616 kWh, 548 kWh and 421 kWh when using the system #1, system #2 and system #3 respectively instead of classical PV chain configuration. For example, the saved energy obtained by using the proposed system #1 is equal to 26.7% of the total PV panel produced energy in a year.

Table 4 resumes the obtained results for the different proposed PV system in Lyon and Sfax Cities.

It is clear that in all the cases that system number 1 gives the best saved energy value compared to other PV system architectures. The advantage of system number 2 and 3 is a very low voltage DC bus is used which is in agreement with actual standards and security limits for people. In the future, many standards should be changed in industrial and domestic applications for a suitable energy management of energy use.

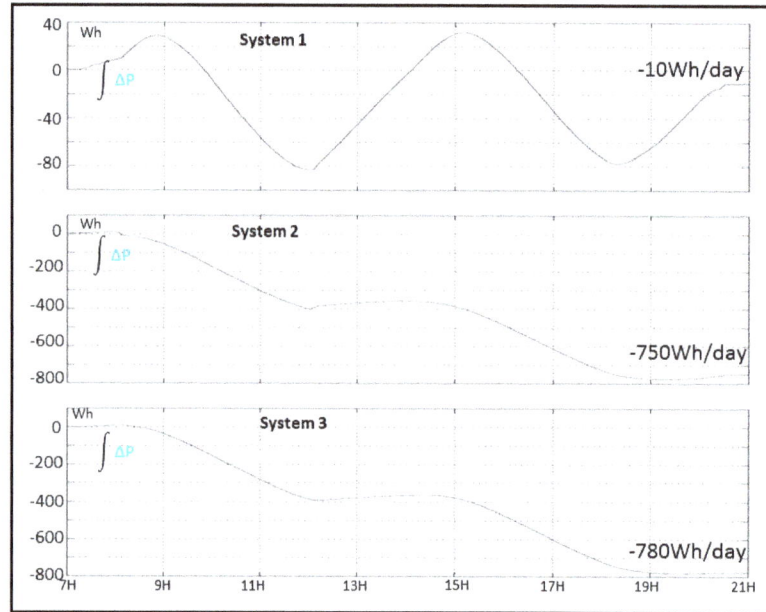

Figure 15. $W(k)$ energy evolutions during a very bad weather day.

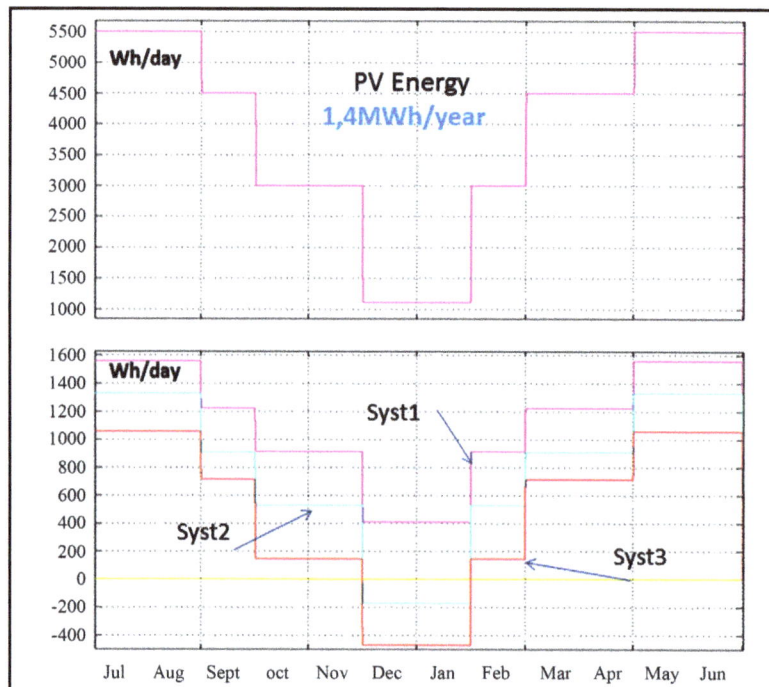

Figure 16. Assumed PV energy and the resulting energy balance $W(k)$ evolutions during a year (Lyon city).

These studies clearly show the importance of the use of a DC bus voltage for DC load supply without using PFC converters and AC plugs. The significant magnitude of the saved energy in these cases should not be ignored. This must incite us to change traditional DC loads input and review the supplying loads modes, especially when PV panel systems are locally used for on-grid applications.

We notice that experimental validation of the obtained results, in this paper, is possible. In our case we had the object to shows the efficiency of the LVDC use by using an accurate model of the different converters. These models are validated experimentally in previous works.

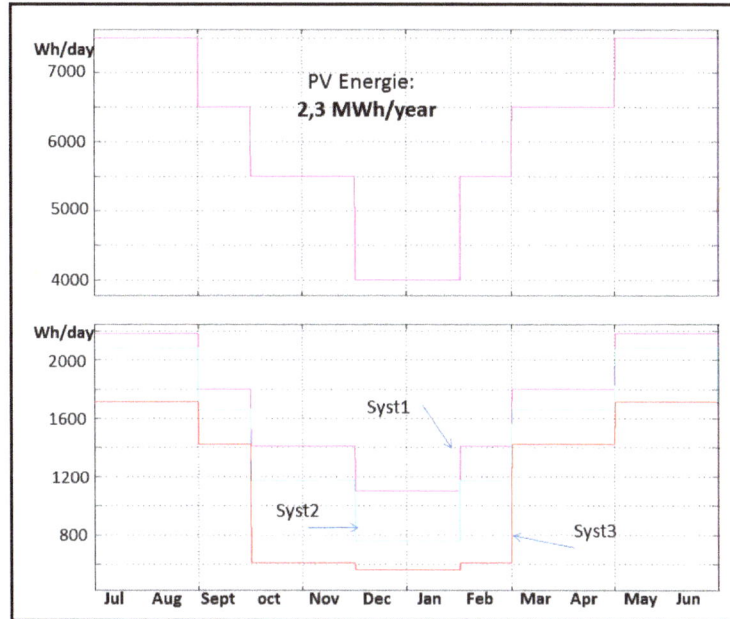

Figure 17. Assumed PV energy and the resulting energy balance W(k) evolutions during a year (Sfax city).

Table 4. Saved energy by the proposed PV chains in the two studied cities.

City	Averaged energy (Wh/day)				Annual Economized energy (kWh)		
	May June July August	Mars April September	February October November	December January	System 1	System 2	System 3
Lyon	5500	4500	3000	1100	404	278	176
Sfax	7500	6500	5500	4000	616	548	421

5. Conclusions

In this paper we have focused on the utilities of the DC bus use, as an input supply for numerous DC loads, in the on-grid PV conversion chains. Classically, in home and office applications, DC loads need alternative to direct current conversion which is not without consequences on the supply network as well as the system sizing and losses. Additional circuits like PFC are often used to improve absorbed current quality. In PV applications, a DC bus is often available and the direct use of this bus to supply DC loads increases considerably the efficiency of the global PV system. To evaluate the efficiency of the different proposed PV chains, non-ideal averaged models of the different converters, have been used. These models are accurate and suitable to complex system studies.

It is registered as an important efficiency of the proposed PV chains, compared to the classical one where supplying DC loads needs standard AC plugs.

Evaluating the annual saved energy of a typical application in two cities, Sfax in Tunisia and Lyon in France, shows that this energy exceeds 25% of the PV generated energy.

The advantage of the proposed studies is more significant when the main consumed energy is in daytime. In this case DC input loads should be used. This is the case of most of the office applications. Probably, the electric system supplies and demands in future will be different from that of today.

Looking for environmental context and economic considerations, close cooperation among many different participants should be deployed to introduce a new standard and to provide suitable power inputs for DC loads.

References

[1] Jaouen, C., Barruel, F. and Multon, B. (2010) Investigation of DC Distribution by Measuring and Modelling Power Supply Devices for Buildings with PV Production.

[2] Par Gilles Notton† et Marc Nuselli. Utilisation Rationnelle de l'énergie et énergies renouvelables, des allies incontestables : Application à une production décentralisée d'electricite photovoltaique. Université de Corse-Centre de Recherches Energie et Systèmes, U.R.A. CNRS 2053, F-20000 AJACCIO, France.

[3] Poissant, Y., Thevenard, D. and Turcotte, D. Mesure en continu de la performance du système photovoltaïque du nunavut arctic college: Neuf années de production fiable d'électricité.

[4] Kaipia, T., Salonen, P., Lassila, J. and Partanen, J. Possibilities of the Low Voltage DC Distribution Systems. Lappeenranta University of Technology, Lappeenranta.

[5] Crudele, D., Key, T., Mansoor, A. and Khan, F. (2004) Commercial and Industrial Applications Getting Ready for Direct Current Power Distribution. *IEE Power Systems Conference and Exposition PSCE*, New York, 10-13 October.

[6] Ammous, A., *et al.* (2002) Developing a PWM-Switch Model Including Semiconductor Device Non-Linéarities. *The European Physical Journal Applied Physics*, **21**, 107-120.

[7] Abid, S., Ammous, K., Morel, H. and Ammous, A. (2007) Advanced Averaged Model of PWM-Switch Operating in Continuous and Discontinuous Conduction Modes. *International Review of Electronic Engineering* (*IREE*), **2**, 544-556.

[8] Fairchild Semiconductor (2004) Application Note 42047. Power Factor Correction Basics. 1-10.

[9] ON Semiconductor (2007) Power Factor Correction Handbook. HBD853/D, Rev. 3.

[10] Coiltronics, Power Factor Correction Application Notes.

[11] Nordman, B., Brown, R. and Marnay, C. (2007) Low-Voltage DC: Prospects and Opportunities for Energy Efficiency, Lawrence Berkeley National Laboratory.

[12] Benalaya, A., Amri, A., Chekirbane, A. and Nmiri, A. Rayonnement Global et Insolation Observés en Tunisie: Potentiel, Relation et Réchauffement Climatique.

13

Performance Evaluation of Listed Companies in New-Energy Automotive Industry

Zexuan Lu, Hongsheng Xia

Management College, Jinan University, Guangzhou, China
Email: luzexuan2008@163.com

Abstract

As one of the national strategic emerging industries, new-energy-automobile industry has been caused people's attention increasingly. Therefore, improving the new energy industry listing corporation performance, not only can alleviate the pressure on energy and the environment, but also can conducive to accelerating the transformation and upgrading of the automotive industry. Moreover, it can foster new economic point of growth and international competitiveness. This essay uses the super efficiency DEA model and Malmquist index method to analyze the performance of new-energy-automotive industry listing corporation. The data is from Year 2011 to 2013. Finally according to the conclusion of the empirical research, this paper made several suggestions to improve the operational efficiency of China's new energy automobile enterprises and promote new-energy-automobile industry.

Keywords

New-Energy-Automobile Industry, Performance Evaluation, Super-Efficiency DEA Model, Malmquist Index

1. Introduction

As the pollution of world becomes more and more serious, the energy crisis is coming. Moreover, along with the sustainable development of low carbon economy, new energy automobile industry has get more attention because of its energy saving and environmental protection. Since our country has added into WTO, the new energy automotive industry in our country has made a significant breakthrough. In 2010, the State Council listed the new energy automotive industry as the strategic emerging industry which leading the development of economic. In 2012, the State Council put forward to seize the opportunity to speed up the cultivation and development of new energy automotive industry, because it can not only alleviate the pressure of energy and environment, but also help to accelerate the transformation and upgrade of the automotive industry. According to some scholar, in the 2014 and 2015, China new energy automobile industry will increase explosively [1]. Therefore, it is significant to develop new energy automotive industry as it can implement the national energy-saving emission reduction strategy and change the mode of economic growth.

However, the development of Chinese new energy automobile industry still faces many challenges, especially the insufficient investment of the core technology field. And it has not yet formed a perfect system of technology innovation. The commercialization of new energy automotive progressed at a slow pace. The new energy automotive products in high cost, the key technology has not developed, the new energy automotive industry is still in the initial stage and other problems that exist in enterprises, such as inadequate investment, local protectionism and other specific issues. Some scholars have found the effect of path dependence unavoidably exists in the process of new energy automotive industry development, And will lead automobile industry to lock in traditional technical and institutional development path [2]. Under this background, it is an important research topic of how to improve operating efficiency of the new energy automobile enterprise and how to enhance the competitiveness of new energy automobile industry.

Based on the above background, firstly, this article starts from view of operating efficiency, and use s super efficiency DEA model to measure the operating efficiency of the new energy automotive listing corporation. Secondly, the further research is to study the dynamic change of total factor productivity (TFP) of new energy automobile industry listing corporation by using the Malmquist index method. The purpose is to find out the factors that influent the operating efficiency of new energy automobile enterprises. Thereby to improve the operating efficiency, enhance the overall competitiveness, and promote the development of new energy automotive industry.

2. Literature Review

With the government's strong support and promotion, the new energy automobile industry is booming. In recent years, scholars at China and abroad have made lots of research on the new energy automotive industry from different angles according to different research purposes. Nevertheless, these researches mainly focus on qualitative research on the development of new energy vehicle technology and industrialization namely. The research of quantitative research on the new energy automobile industry and enterprises is rare, needless to say the research of new energy automotive industry business performance of the company.

The research of foreign scholars on automobile enterprise efficiency and influence factors is earlier. Seema-Sharma (2006) [3] studied India automobile industry by using the method of total factor productivity index analysis, and further explored the influence factors of automobile industry efficiency. Papahristodoulou (1997) studied the production efficiency of German car enterprises using DEA model [4]. Many scholars used DEA model to analyze the traditional automobile industry or the automotive component manufacturing production efficiency, and put forward some suggestions and countermeasures to improve the production efficiency of automobile enterprises (R. J. Orsato and P. Wells, 2007; Saranga, 2009; Nandy, 2011) [5]-[7].

Research on automobile enterprises efficiency of Chinese scholars was later. Yong Wang (2010) studied the efficiency of automobile enterprises by using the DEA method., and come to the conclusion that the total factor productivity of auto mobile industry listing corporation was in an upward trend between 2001 to 2008 [8]. In addition, some scholars found that technical efficiency contributed greatly on total factor productivity of China's automobile enterprises (Zenghui Li, 2012, Xianjin Wu, 2011) [9] [10]. What had discussed above were the traditional research in the automotive industry, the researchs for new energy automobile industry were quite few, not to mention about the study on the performance of new energy automobile enterprises. Recently, the study mostly focued on its industrial policy, enterprise strategy and regional development, For instance, Yang Ping and Ke Chuan Yi (2011) [11] made SWOT analysis of new energy vehicles. They concluded that the new energy automotive industry should establish industry alliance, integrate the advantage resources to the development of new energy vehicles in China; increase investment in research and development, increase policy subsidies, improve the supporting facilities, strengthen international cooperation and so on. There were scholars also making qualitative research on new energy vehicles, Jing Ruan (2010) [12] used comparative analysis and expert scoring method on the basis of evaluation index system of new energy vehicle, and further used fuzzy comprehensive evaluation method to judge comparison of the advantages and disadvantages of its development, finally obtained the long-term new energy technology development route. In addition, some scholars established input-output index system to study the technology innovation efficiency of new energy automotive industry (Wang

Wei, 2013) [13] [14]. The business model is the key to the industrialization of new energy vehicles. In order to construct the need for new energy vehicles business model. We should entail the explicit priorities, macro-guidance and coordinating the related interests [15]. As the battery is the critical part of new energy vehicle, some scholars studied research on the performance and service life of batteries membrane of new energy auto-motive [16] [17]. Moreover, many scholars focus on the analysis of policies for new energy vehicle by making researches on the problems and challenges in China's new energy vehicle industry [18]-[21].

By sorting out the early scholars study on the new energy automotive industry we found that, from the qualit-ative perspective, the prior analysis of new energy vehicles is lack of objective data of enterprise input and out-put. Therefore, this paper will analyze the static and dynamic efficiency of new energy automobile industry list-ing corporation by using the method of super efficiency DEA model and Malmquist index. Thus this paper will put forward some reasonable suggestions to improve the performance the new energy automotive industry and enhance the core competence of the new energy automotive list companies.

3. Research Methods and Data Processing

By reading literature we found that scholars who studied on the management efficiency of enterprises were more likely to adopt the DEA model, however most of the existing DEA model has short comings: Firstly, if there are more than one effective DMU it cannot distinguish them effectively. Secondly, DEA model cannot study the dynamic changes unit efficiency. Based on this, this paper uses the method of super efficiency DEA to compare of multiple efficiency DMU, and uses the Malmquist index approach to analyze the dynamic efficiency of total factor productivity of new energy automobile industry listing corporation.

3.1. The Super Efficiency DEA Model

The super efficiency DEA model was composed by Andersen and Pelersen in 1993 which based on the DEA model. The model can compare the efficiency between the various effective DMUs. In this paper, we choose super efficiency DEA model to study the performance of the listed new energy automobile company in China. The model is input oriented.

Suppose that there are n kind of companies, each company $j\,(j=1,2,\cdots,n)$ inputs m kinds of factors of production $X_{ij}\,(i=1,2,\cdots,m)$ and produces s kind of output $Y_{rj}\,(r=1,2,\cdots,s)$. Expressions of super efficiency DEA model are showed as follows:

$$\min \theta_0^{\text{super}}$$

$$s.t. \quad \sum_{\substack{j=1\\j\neq0}}^{n} \lambda_j X_{ij} + S_i^- = \theta_0^{\text{super}} X_{i0} \tag{1}$$

$$\sum_{\substack{j=1\\j\neq0}}^{n} \lambda_j Y_{rj} - S_r^+ = Y_{r0} \tag{2}$$

$$\sum_{j\neq0} \lambda_j = 1$$

$$\lambda_j, S_i^-, S_r^+ \geq 0, j \neq 0$$

Equation (1) is the expression of super efficiency DEA model, and θ_0^{super} is the super efficiency values of DMU$_0$. In equation (2), S_i^-, S_r^+ are the slack variables.

3.2. Malmquist Productivity Index

The Malmquist index was developed by Swedish economist Malmquist in 1953. It based on the distance func-tion which regarded as a kind of consumption index. Caves and Fare applied it to calculate the index of multiple inputs and outputs of the total productivity change. Total factor productivity (TFP) refers to the comprehensive productivity. It is widely applied to the enterprise productivity. Study on the window period of S to T, of varia-ble returns to scale, output oriented Malmquist index can be defined as:

$$m(ys,xs,yt,xt) = \left[\frac{d_s^C(y_t,x_t)d_s^C(y_s,x_s)}{d_t^C(y_t,x_t)d_t^C(y_s,x_s)} \right]^{1/2} \left[\frac{d_s^C(y_t,x_t)d_s^C(y_s,x_s)}{d_t^V(y_t,x_t)d_t^C(y_s,x_s)} \right] \left[\frac{d_s^V(y_t,x_t)}{d_t^V(y_s,x_s)} \right] \tag{3}$$

Equation (3) is the expression of Malmquist index. $d^C(y,x)$ is the distance function that under the invariable situation of returns to scale, $d^v(y,x)$ is the distance function of variable returns to scale. The expression for the first term is "technological progress". The second is scale efficiency change; The third is "pure technical efficiency change". Total factor productivity (TFP) can be decomposed into technological progress, pure technical efficiency and scale efficiency change. Technical efficiency change is further decomposed into pure efficiency change and scale efficiency change. When the technical efficiency change is greater than 1, it indicates that T period is much more close to the production surface relative to the S period, explained that the enterprise has improved the efficiency due to the effective management and decision. Otherwise is invalid. The same as the technical efficiency change. When it is greater than 1, indicates that the enterprise improve its efficiency due to technological progress, and vice versa. When total factor productivity is greater than 1, it shows that total factor productivity is increased, and vice versa is reduced.

3.3. Choosing Evaluation Indexes

The key research and analysis of DEA model is to select the input and output indicators reasonably. This paper follows some principles when select the indicators, such as the index should reflect the competitiveness level of DMU objectively. The data can be obtained, Moreover this paper selects the following indicators based on literature and experts' consultations.

1) The input index

For enterprises, investment of resources can be divided into human resources, material resources and financial resources. Therefore, the input index can define into the number of employees ($X1$), total assets ($X2$), the main business costs ($X3$). Every employee in the enterprise carries out their duties to bring the enterprise benefit. The total assets of the enterprise including the fixed assets, current assets, intangible assets and other material resources, The total assets is the basis of enterprise management and development. Main business cost refers to all the expenses occurred by company's main business in the production and business operation process, including the production of raw materials, the sales cost and other costs. This index reflects the enterprise's financial resources from the angle of investment cost.

2) The output index

Standing in the perspective of stakeholders, the enterprises output can measured for its contribution to the stakeholders. Stakeholders of the firm are mainly divided into shareholders, employees, creditors and the government. Accordingly, the output indicators divided into net profit ($Y1$), employee compensation ($Y2$), interest expenses ($Y3$) and income tax expense ($Y4$).

3.4. Research Object and the Data Source

The new energy automotive industry chain involves in many enterprises. The sample comes from the new energy automotive industry plate in Shanghai and Shenzhen stock market. The sample also refers to the "2013 energy-saving and new energy vehicles Year book". The sample excludes ST companies, a serious lack of data and incomplete disclosure of the company. Finally the paper selects 36 companies as the research samples. All the data come from financial statement of these 36 listing corporation from 2011-2013.

4. Empirical Analysis

Based on the input oriented DEA-BCC model, this paper uses DEAP 2.1 software to measure the output management efficiency of 36 new energy automobile industry listing corporations in 2013. The article also figures out the technical efficiency, the pure technical efficiency and scale efficiency value of each DMU. In DEA evaluation, while technical efficiency is equal to 1, the production of the enterprise is efficient, and we called it DEA effectively. Otherwise if the technical efficiency is less than 1, the production of the enterprise is inefficient. Due to there are more than one efficiency value of new energy mobile listing corporation, in order to calculate the efficiency value further ,this paper uses EMS software to calculates the super efficiency DEA on DMU. The DEA efficiency and calculation efficiency of super DEA are as shown in **Table 1**.

4.1. Analysis of DEA Model

From the overall perspective, in 2013, 9 of the 36 new energy automotive listed companies are DEA effective, it

Table 1. The DEA efficiency super DEA efficiency of new energy automotive industry listing corporation.

DMU	Overall Efficiency	Pure Technical Efficiency	Scale Efficiency	Scale	The Super Efficiency (%)	The Efficiency Rating
Inovance Technology	1	1	1	-	2.554	1
Wanxiang Qianchao	1	1	1	-	2.318	2
Xiamen Tungsten	0.947	1	0.947	drs	2.098	3
Nari Technology	1	1	1	-	2.06	4
BYD	0.554	1	0.554	drs	2.001	5
Rongxin Power Electronic	1	1	1	-	1.834	6
Auto Electric Power Plant	0.452	1	0.452	irs	1.29	7
Zhongheng Electric	0.601	1	0.601	irs	1.238	8
Xiamen Faratronic	1	1	1	-	1.226	9
Ningbo Yunsheng	1	1	1	-	1.1	10
Sieyuan Electric	0.924	1	0.924	drs	1.017	11
Hainan Sundiro Holding	0.92	1	0.92	irs	1	12
Qinghai Salt Lake Industry Group	1	1	1	-	1	13
Shanghai General Motors	1	1	1	-	1	14
Inner Mongolia Baotou Steel Rare-Earth	0.744	0.993	0.749	drs	0.993	15
Anhui Jianghuai Automobile	0.851	0.99	0.859	drs	0.99	16
Zhejiang Founder Motor	0.154	0.948	0.163	irs	0.948	17
Jiangsu Guotai	0.927	0.94	0.985	irs	0.94	18
Anyuan Coal Industry Group	0.737	0.887	0.831	drs	0.887	19
Ningbo Joyson Electronic	0.797	0.822	0.97	drs	0.822	20
Shenzhen Clou Electronices	0.688	0.797	0.864	irs	0.797	21
Zhongshan Broad-Ocean Motor	0.762	0.791	0.963	irs	0.791	22
Henan Senyuan	0.785	0.91	0.863	irs	0.765	23
XJ Electric	0.661	0.722	0.915	drs	0.722	24
Xiamen King Long Motor Group	0.706	0.773	0.913	drs	0.722	25
Nantong Jianghai Capacitor	0.526	0.707	0.745	irs	0.707	26
Jiangxi Ganfeng Lithium	0.415	0.815	0.509	irs	0.656	27
Jiangxi Special Electric Motor	0.257	0.59	0.436	irs	0.587	28
Sino-Platinum Metals	1	1	1	-	0.581	29
Ningbo Shanshan	0.519	0.536	0.968	irs	0.536	30
Beiqi Foton Motor	0.429	0.535	0.802	drs	0.524	31
Dongfeng Automobile	0.742	0.816	0.909	drs	0.456	32
ZhongTong Bus & Holding	0.309	0.42	0.736	irs	0.42	33
Shanghai Tongji Science & Technology Industrial	0.572	0.783	0.73	irs	0.313	34
Lanzhou Greatwall Electrical	0.18	0.292	0.616	irs	0.292	35
Zhuzhou Times New Materials	0.26	0.346	0.752	irs	0.233	36
Mean	0.706	0.845	0.824			

accounts for 25% of the total and the average comprehensive performance scores 0.706, indicating that these 9 decision units are in the efficient production frontier. The remaining 27 companies are non DEA effective, indicating this 27 enterprises resources have not reached the optimal configuration. They can still improve and optimize the allocation of resources. The least level of comprehensive technical efficiency is the Great Wall electrician which only scores 0.18. The overall result shows that the level of resource configuration of new energy automotive listing corporation is not high.

According to the data in **Table 1**, we can divide the new energy automotive listing corporation into three categories:

The first class is the pure technical efficiency and the scale efficiency are less than 1. It means that the scale and efficiency of the company are invalid. There are Baotou Steel Rare Earth, Jianghuai Automobile, founder motor, Xu Ji electric, Jinlong automobile, Zhongtong bus, Tongji Science and technology company. On the one hand, the allocation of resource of these companies is not rational. On the other hand, the level of management efficiency of these companies is not high, the scale of operation needs to be improved. In order to improve the operating efficiency, the companies need to expand the scale.

The second category is the pure technical efficiency scoring 1 but the scale efficiency less than 1. There are 6 companies belong to this category. The allocation of resource of these companies is relatively efficient. Xiamen tungsten industry, BYD, Siyuan electric listing companies are in the decreasing returns to scale stage, indicating that the allocation of resources of these companies is not reasonable. The outspread rate is too rapid and resource utilization rate is too low. Therefore, these companies need to improve efficiency by downsizing properly. Auto Electric Power Plant, Hainan Sundiro Holding are at the stage of increasing returns to scale, it improves that the development trend of the enterprise is good, these companies can increase the output by increasing the human capital and other production factors, so as to improve the performance of the company.

The third category is the scale efficiency and pure technical efficiency scoring 1. It means that the integrated technical efficiency value is 1, and the scale and efficiency of these companies are effective. There are 8 companies belong to this type. The size of configuration and the factor of production resources are in the most efficient state. In order to calculate and distinguish these effective decision unit values further. This paper uses super efficiency DEA model to analyze the effective decision unit values.

4.2. Super Efficiency DEA Results and Analysis

The super efficiency DEA model can further compare the multiple effective decision unit, which will calculate the grasp of the new energy automotive industry listing corporation operating efficiency accurately. In this paper, we use EMS 1.3 software to calculate super efficiency DEA. Seen from **Table 1**, the rank of comprehensive efficiency of super DEA is almost same as the traditional efficiency DEA. The top few are Inovance Technology, Wanxiang Qianchao, Xiamen Tungsten Nari Technology, BYD, Rongxin Power Electronic, Auto Electric Power Plant, Zhongheng Electric, Xiamen Faratronic, Ningbo Yunsheng. The efficiency of these companies is relatively high. The main reason is that these companies mainly focused on research and development. As a strategic emerging industry, research and development ability are the key to obtain the competitive. For example, Huichuan technology company has the research and development advantages in the new energy vehicle field to meet the customer's demand. Therefore, it had got the key customers order in 2013. As "China's Tesla", the development of Wanxiang Qianchao is also very optimistic. After acquisition the part of battery technology of Leiden energy company, it is expected to strengthen technical reserves in the automotive industry. Xiamen tungsten as the upstream supply chain enterprises of new energy automobile industry is the leading enterprises producing Ni MH battery hydrogen storage powder in domestic. Once the new energy automotive industry have developed, Xiamen tungsten industry will expand rapidly. In addition, as the charging station leader, Nari, absolutely is the beneficiaries of the development of new energy automotive industry.

4.3. Malmquist Index

On the basis of DEA research to obtain static operating efficiency the new energy automobile industry listing corporation, this paper will use DEA 2.1 software to calculation the Malmquist productivity index of the new energy automotive industry listing corporation from 2011 to 2013. The calculated results are shown in **Table 2**. The analysis of **Table 2** shows as followed.

First, from 2011 to 2013, the average total factor productivity of new energy automobile industry listing cor-

Table 2. Malmquist productivity index of new energy automotive industry listing corporation from 2011 to 2013.

DMU	Technical Efficiency	Technological Progress	Pure Technical Efficiency	Scale Efficiency	Total Factor Productivity
Shenzhen Clou Electronices	1.489	1.1	1.53	0.973	1.637
Rongxin Power Electronic	1	1.444	1	1	1.444
Wanxiang Qianchao	1.073	1.272	1	1.073	1.365
ZhongTong Bus & Holding	1.543	0.851	1.474	1.047	1.314
XJ Electric	1.642	0.798	1.748	0.939	1.311
Anhui Jianghuai Automobile	1.276	0.944	1.155	1.105	1.205
Zhongheng Electric	1.722	0.656	1	1.722	1.13
Dongfeng Automobile	1.325	0.852	1.24	1.069	1.13
Lanzhou Greatwall Electrical	1.123	0.97	1.126	0.997	1.089
Nantong Jianghai Capacitor	1.66	0.653	1.383	1.2	1.084
Inovance Technology	1	1.074	1	1	1.074
Ningbo Joyson Electronic	1.336	0.785	1.129	1.184	1.049
BYD	0.966	1.083	1	0.966	1.046
Zhongshan Broad-Ocean Motor	1.105	0.922	1.096	1.009	1.018
Jiangxi Special Electric Motor	1.384	0.731	1.122	1.233	1.011
Henan Senyuan	1.272	0.791	1.116	1.14	1.006
Sieyuan Electric	0.961	1.034	1	0.961	0.993
Xiamen King Long Motor Group	1.033	0.951	0.946	1.092	0.982
Nari Technology	1.264	0.773	1.224	1.032	0.977
Xiamen Faratronic	1.15	0.85	1	1.15	0.977
Qinghai Salt Lake Industry Group	1.045	0.931	1	1.045	0.973
Beiqi Foton Motor	0.955	1.015	0.921	1.037	0.969
Auto Electric Power Plant	0.825	1.173	1	0.825	0.968
Sino-Platinum Metals	1.112	0.871	1	1.112	0.968
Jiangxi Ganfeng Lithium	1.317	0.733	0.917	1.436	0.965
Ningbo Shanshan	0.72	1.266	0.745	0.966	0.912
Ningbo Yunsheng	1	0.895	1	1	0.895
Hainan Sundiro Holding	0.959	0.926	1	0.959	0.889
Xiamen Tungsten	1.127	0.762	1.078	1.045	0.858
Anyuan Coal Industry Group	0.987	0.838	1.021	0.967	0.826
Jiangsu Guotai	0.963	0.802	0.963	1	0.772
Shanghai Tongji Science & Technology Industrial	1.241	0.585	1.144	1.085	0.726
Zhejiang Founder Motor	0.737	0.887	0.606	1.216	0.654
Inner Mongolia Baotou Steel Rare-Earth	0.862	0.554	0.997	0.865	0.477
Shanghai General Motors	1	0.435	1	1	0.435
Zhuzhou Times New Materials	0.51	0.852	0.51	0.999	0.434
Mean	1.098	0.867	1.037	1.058	0.951

poration scores 0.951. It shows a negative growth of total factor productivity from 2011 to 2013. Total factor productivity of these three years has decreased by 4.9%. By the same token, technical efficiency increased by 9.8%, but technical progress rate decreased by 13.3%. It suggests that the reduction of total factor productivity is the main reason for the decline of the rate of technological progress. Technological progress has become the key factor of further development of new energy automotive industry.

Second, there are 16 of 36 new energy automobile industry listing corporations achieve positive growth in total factor productivity. According to the relevant theory, when total factor productivity index falls in the interval [1.025, + infinity], it is considered that Malmquist productivity index has obviously risen. When the total factor productivity index falls in the range of [0, 0.975], it is considered that the index has obviously decreased. While the total factor productivity index falls in the interval (0.975, 1.025), the index change is relatively stable. Data showed the total factor productivity of following companies rises obviously: Clou Electronices, Rongxin Power, Wanxiang Qianchao, ZhongTong Bus & Holding, XJ Electric, Jianghuai Automobile, Zhongheng Electric, Dongfeng Automobile, Lanzhou Greatwall Electrical, Nantong Jianghai Capacitor, Inovance Technology, Ningbo Joyson Electronic, BYD. However, the total factor productivity of Jiangsu Guotai, Tongji Science & Technology, Founder Motor declines obviously. This shows that the development of new energy automotive industry listing corporation is not balanced.

Third, from the decision-making unit, the total factor productivity of CLOU electronics grows fastest. It is 63.7%. It all comes down to the technical efficiency increase (increased by 48.9%), while the technical efficiency increase is mainly due to pure technical efficiency growth (increased by 53%). On the whole, there are 16 listing corporation's total factor productivity growth rate greater than 1. Relative to other listing corporation, the decline of technology progress of these 16 companies is narrow. Therefore, the technological progress rate directly determines the level of total factor productivity value and the change value of new energy automotive industry.

5. Conclusions and Recommendations

5.1. Conclusions

This article carries on the analysis and evaluation of operating performance of 36 new energy automobile industry listing corporation on the base on super efficiency DEA and Malmquist index method.

The empirical results show that there are only 9 listing corporations reached DEA effectively among 36 new energy automobile industry listing corporations. From the overall perspective, the performance of new energy automotive industry is general. From the super efficiency value point of view, even DEA effective companies still have to improve the efficiency. From the Malmquist total factor productivity index, the average growth of total factor productivity of new energy auto industry listing corporation is −4.9% from 2011 to 2013, the study finds that the rate of technological progress is an important cause of the decline of total factor productivity. The result shows that in new energy automobile industry, the productivity depends more on technical efficiency to some extent, while ignoring the importance of technological progress, technological progress has become the key factor restricting the new energy automotive industry for further development.

5.2. Recommendations

Although the new energy automotive industry in China has made positive progress, as a strategic emerging industry, the listing corporations are at the initial stage overall. In order to enhance the operational efficiency of new energy automobile enterprises, promote the long-term development of new energy automotive industry. Based on the empirical analysis results, this paper proposes the following suggestions from government, industry and enterprises perspective.

1) The government should strengthen regulation policy, establish and improve the system of technology innovation. As the enterprises input enormous cost in R & D and market cultivation. On the one hand, the government should establish policy of the finance and taxation system reasonably, so that to reduce financial burdens on enterprises and guarantee the development of new energy vehicles. On the other hand, the government should establish the effective investment and financing system to the integrate capital resource, technical resource and other parties resources.

2) The development of the new energy automobile industry is inseparable from the charging network of urban

infrastructure, the government should introduce relevant policies to encourage the construction of infrastructure of new energy automobile and reduce the new energy input costs. Moreover, the government should improve the efficiency of the whole management of new energy automobile companies, so that to enhance the overall competitiveness of new energy automobile industry and promote the development of new energy automotive industry.

3) From the evaluation results, Firstly, the new energy automobile enterprises should focus on technological progress, they have better increase technology research and development investment, and strengthen the independent innovation and technological upgrading. In addition, the companies should improve the efficiency of input and output of new energy automobile enterprises. Secondly, it is significant to cultivate high quality talents and scientific research team. What is more, establishing the mechanism of introducing talents is also critical. One of the feasible and effective way is to cooperate with of university in educating high-quality talents.

4) Last but not the least, the new energy automotive industry can establish industry alliance and integrate the advantage resources to strengthen the connection and cooperation of the industry chain. Only the whole new energy automobile industry chain makes balanced development can the new energy automobile enterprises cultivate core competence and promote the comprehensive development of the new energy automotive industry.

References

[1] Jia, L.J. and Lv, R.Z. (2014) New Energy Vehicle Takes Great Leap Forward. *Automobile & Parts*, **14**, 32-36.

[2] Zhang, G.Q. and Zhang, X. (2014) Path Dependence and Solutions Faced by Development of New-Energy Automobile Industry. *Journal of Industrial Technological Economics*, **2**, 75-80.

[3] Sharma, S. (2004) A Study on Productivity Performance of Indian Automobile Industry: Growth Accounting Analysis. *APPC* 2004 *Conference Program*, Hanoi, Vietnam, 25-31 October 2004.

[4] Papahristodoulou, C. (1997) A DEA Model to Evaluate Car Efficiency. *Applied Economics*, **29**, 1493-1508. http://dx.doi.org/10.1080/000368497326327

[5] Orsato, R.J. and Wells, P.U. (2007) The Rise and Demise of the Automobile Industry. *Journal of Cleaner Production*, **15**, 994-1006. http://dx.doi.org/10.1016/j.jclepro.2006.05.019

[6] Saranga, H. (2009) The Indian Auto Component Industry—Estimation of Operational Efficiency and Its Determinants Using DEA. *European Journal of Operational Research*, **196**, 707-718. http://dx.doi.org/10.1016/j.ejor.2008.03.045

[7] Nandy, D. (2011) Efficiency Study of Indian Automobile Companies Using DEA Technique: A Case Study of Select Companies. *The IUP Journal of Operations Management*, **10**, 39-50.

[8] Wang, Y. and Ji, Y. (2010) Study on Efficiency of Chinese Automobile Industry Listing Corporation. *Commercial and Research*, **12**, 97-101.

[9] Li, Z.H. (2012) Research on the Construction and Innovation Path Selection Evaluation of Multidimensional Model of China Automobile Industry. Ph.D. Thesis, Wuhan University of Technology, Wuhan.

[10] Wu, X.J. and Chen, X.L. (2011) An Empirical Analysis of Total Factor Productivity and the Impact of China's Automobile Industry. *Research on Financial and Economic Issues*, **3**, 41-45.

[11] Yang, P. and Ke, C.Y. (2011) SWOT Analysis China's Development of New Energy Vehicles in the Post Crisis Era. *Exploration of Economic Problems*, **3**, 18-23.

[12] Ru, X.J. (2010) Research on Comprehensive Evaluation of New Energy Automotive Technology Economy and Its Development Strategy. Ph.D. Thesis, Wuhan University of Technology, Wuhan.

[13] Wang, W. (2013) Research on Innovation Efficiency of New Energy Automobile Enterprise. Ph.D. Thesis, Harbin Institute of Technology, Harbin.

[14] Zhang, G.W. and Zeng, H.Y. (2014) Study on the Factors of New Energy Automobile Production Efficiency and Effects Based on DEA-Tobit Model. *Industrial technology economy*, **3**, 130-137.

[15] Quo, Q.F. (2014) Research on the Business Model of Development of the New Energy Automobile Industry. *Special Zone Economy*, **7**, 45-47.

[16] Li, Y., Song, J. and Yang, J. (2012) Progress in Research on the Performance and Service Life of Batteries Membrane of New Energy Automotive. *Chinese Science Bulletin*, **57**, 4153-4159. http://dx.doi.org/10.1007/s11434-012-5448-9

[17] Li, Y., Song, J. and Yang, J. (2014) A Review on Structure Model and Energy System Design of Lithium-Ion Battery in Renewable Energy Vehicle. *Renewable and Sustainable Energy Reviews*, **37**, 627-633. http://dx.doi.org/10.1016/j.rser.2014.05.059

[18] Liuqin, C. and Bing, X. (2012) China's New Energy Vehicle Industry: Problems and Challenges. *Electric (English*

version), **33**.

[19] Gong, H., Wang, M.Q. and Wang, H. (2013) New Energy Vehicles in China: Policies, Demonstration, and Progress. *Mitigation and Adaptation Strategies for Global Change*, **18**, 207-228. http://dx.doi.org/10.1007/s11027-012-9358-6

[20] Gu, L.Z. and Shao, Y.F. (2014) The Analysis of Innovation Policies for New Energy Vehicle Technology. *Studies in Sociology of Science*, **5**, 147-151.

[21] Xi, X. and Zhao, J.Y. (2014) Breaking the Existing Regional Technological Regime with Effective Policy-Making: A Case Study on New-Energy Vehicle Manufacturing in Jilin Province. 2014 *IEEE International Conference on Management Science & Engineering* (*ICMSE*), Helsinki, 17-19 August 2014, 2102-2109.

14

Analysis of Technical Energy Conservation Potential of China's Energy Consumption Sectors

Xi Yang[1,2], Qing Tong[1], Xunmin Ou[1,2*]

[1]Institute of Energy, Environment and Economy, Tsinghua University, Beijing, China
[2]China Automotive Energy Research Center, Tsinghua University, Beijing, China
Email: ouxm@tsinghua.edu.cn

Abstract

It is necessary for China to refocus its energy conservation effort from the industrial sector (field) to all three sectors simultaneously, *i.e.* industry, construction and transport. In addition, it should also make significant effort for conserving energy on general technical equipment that are used in large quantities and for a variety of applications. Therefore, there is a need to integrate industrial, construction and transport sectors, *i.e.* the integration between key technologies and widely used technologies, between hard and soft management, between energy-saving technologies and comprehensive resource utilization technologies. According to estimates, if China's energy consuming sectors adopted appropriate energy-saving technologies, total energy-savings (using 2010 as the baseline) would be 200 million, 450 million, 650 million and 800 million tons of standard coal in 2015, 2020, 2025 and 2030, respectively.

Keywords

Energy Conservation, Technology Energy-Saving, Policy Analysis, China

1. Introduction

"Energy Reduction Revolution" is mainly realized via the following two approaches: by reducing the consumption of energy by services and improving energy utilization efficiency (*i.e.* using less energy for a similar output). The latter is generally called technical energy conservation, *i.e.* continuously improving energy utilization efficiency of existing and new energy facilities (or equipment), and reducing the energy consumption per unit out-

---

*Corresponding author.

put of energy services. Technical energy conservation is further divided into progressive and fundamental types. Thus, in addition to gradual increase of energy efficiency through incremental innovation of the main technologies in service, additional exploration of fundamental technological innovations should be undertaken to achieve breakthroughs in energy efficiency.

2. Estimation of Technical Energy Conservation Potential by Sector

Energy conservation applies to every sector of the national economy and social life, and can be realized through a variety of approaches and measures. As for the methodology, prioritizing key fields and promoting effective measures can accomplish the greatest performance. Key energy conservation sectors must adhere to the idea of "doing two jobs at once"; that is, paying attention to areas consuming high amounts of energy, and to general technical equipment that are commonly used in large quantities. In the past, all energy conservation work revolved around the industrial sector as it had the highest energy consumption. However, in the future, the energy conservation potential of the industrial sector will wane, and will shift to the consumption structure of people "living" and "traveling". The energy consumption in residential & commercial real estate and the transport & logistics sectors will grow rapidly (**Figure 1**) [1] [2]. Therefore, it is necessary to refocus energy conservation from the industrial sector (field) to all three sectors, *i.e.* industry, construction and transport, and to increase the energy conserving effort from general technical equipment that are used in large quantities and for a variety of applications.

Special attention should be paid to the industrial, construction (including tertiary industries and household consumption) and transport sectors—the main energy consuming sectors of China (**Table 1**). Industrial energy usage accounts for over 70% of China's energy consumption; thus it is a key sector for energy conservation. In addition, accelerated urbanization and motorization has led to a rapid growth in energy consumption within China's construction and transport sectors.

Standards for energy conservation are set for the purpose of energy saving, and used as the fundamental basis to control energy consumption from source. Hence, energy conservation should be considered as a focus of China's energy policies, and relevant industrial standards, rules and regulations related to energy conservation should constantly be developed and improved.

2.1. Industrial Sector

Though the industrial sector saw great accomplishments in energy conservation, a number of issues still need to be addressed: energy efficiency needs to be improved, the promotion of energy-saving technologies and equipment is not sufficient, supporting regulations and system standards have to be perfected and so on. Currently, there is a gap in energy consumption per unit output value and per unit production between the Chinese industrial sector and that of developed countries. Thus, there is considerable energy conservation potential.

The comparison between the unit energy consumption of the main energy-intensive products between China

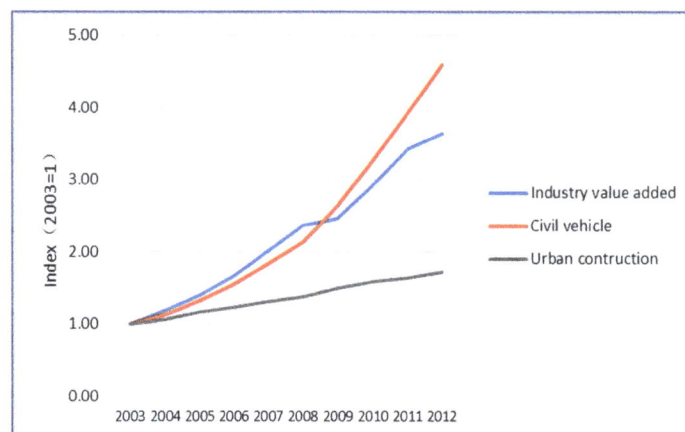

Figure 1. Growth of industrial, value-added civil vehicle and urban building area in China (2003-2012).

Table 1. Proportion of final energy consumption by sector[a] (%).

Sector	2005	2010	2011
Primary industry	2.6	2.0	1.9
Industrial sector	71.5	71.1	70.8
Construction	1.4	1.9	1.7
Transport	7.8	8.0	8.2
Wholesale, retail, accommodation, and catering sectors	2.1	2.1	2.2
Other industries	3.9	4.2	4.4
Household consumption	10.7	10.6	10.7

[a]Data Source: Yearbook of Energy Statistical Records. According to internationally accepted statistical standards. The energy consumption in transport includes both operational transport and transport in other sectors; China Energy Research Society quoted IEA data in the Annual Report on China's Energy Development (2013), and said that in 2010 energy consumption in China's transport sector accounted for 12.1% of final energy consumption, which is below the global average of 23.9%; Construction energy consumption is the energy consumed in non-production construction, *i.e.* in civil construction (residential and commercial buildings).

and foreign countries shows that the energy efficiency of some industrial sectors of China still needs improvement. The emphasis should be on technology development and their application, where required. For instance, as described in the appendix of China Energy Statistics Yearbook (2013), energy consumption per ton of steel, comprehensive energy consumption per ton of cement, and per ton of paper and paperboard in China is 10% - 40% higher than Japan. China uses naphtha and coal as the main raw material to produce ethylene and synthesize ammonia, respectively. Thus the corresponding overall energy consumption is about 50% higher than the international advanced level (the Middle East uses ethane to produce ethylene, and USA uses natural gas to synthesize ammonia).

In terms of standards, it is necessary to accelerate the formulation (revision) of standards for limited energy consumption per unit product (process); increase the implementation of energy efficiency standards in industrial equipment, household appliances and lighting, information and communications and other fields; encourage the development of stringent local standards for unit consumption and energy efficiency; promote international coordination and unification of energy conservation standards for industry; strengthen the certification of energy-saving products and the establishment of inspection capability; expand the scope of energy efficiency labeling of energy-saving products and increase government procurement of energy-saving products.

The Institute of Energy, Environment and Economy, Tsinghua University, concluded that the amount of energy-saving over, a fixed period within each five-year plan period from 2010-2025, will be approximately 100 million tons of standard coal following the implementation of key industrial energy-saving technologies.

2.2. Construction Sector

Energy saving in construction refers to energy conservation and consumption reduction during the entire life cycle of buildings. Due to limited statistical standards and methods, no comprehensive energy consumption data of the entire building sector is available in China at the moment. Relevant studies suggest that the energy consumption of buildings makes up about 20% of total energy consumption across China. Buildings are the main source of energy consumption within the tertiary industry, wholesale and retail, accommodation and catering sectors and two other main industries as well as households. The Building Energy Research Centre, Tsinghua University, estimated that the current total energy consumption in building operations is approximately 680 million tons of standard coal in China. However, with future increases in construction the energy consumption in buildings can still be controlled at 840 million tons of standard coal by adopting a series of technologies, standards and management measures.

China has achieved remarkable results in energy saving in buildings; however, some problems still need to be addressed. These include, unbalanced progress of energy conservation in different regions; differing construction stages and types; lack of scientific guidance to select energy-saving technologies; technical difficulties in saving energy in buildings as per local conditions; inadequate incentives and restrictions; insufficient awareness of energy-saving within enterprises and individuals; poor basic information about energy-saving strategies in building, amongst other areas.

Over the past decade, a number of energy-saving technologies have been widely applied in new and renovated energy-saving buildings, with most showing improved results. In the future a number of increasingly competitive technologies with improved energy conservation potential and emission reduction must be promoted. If energy-saving and emission reduction measures are effectively implemented, approximately 70 - 80 million tons of standard coal equivalents can be saved, over each five-year plan period from 2010-2020, within the building sector.

There is no uniform implementation of energy conservation standards for new buildings. On the whole, the energy-saving standards promulgated for the building sector, during the 11th Five-Year Plan, were 50% of the consumption level between 1980-1981 (the 50%-standard), which was gradually increased to 65% at the end of the 11th Five-Year Plan. This indicates low levels of energy efficiency standards.

2.3. Transportation Sector

The existing statistical system of China, which contains the energy consumption statistics for the communication and transportation sectors, only includes enterprises engaged in social operations. The system does not include data for the energy consumption of industrial enterprises, privately owned vehicles of enterprises and public units, and private cars. As per general international standards it is estimated that the actual energy consumption of China's transportation sector is far more than current statistics. Research estimates that China's transportation sector accounts for 10% to 15% of the total national energy consumption. Experience from developed countries suggests that on entering a steady state of urbanization, communication and transportation will reach 30% to 40% of the country's total energy consumption.

The transportation sector is one of the fastest growing areas of energy consumption. While fully implementing the prioritized developmental strategy of public transport, developing intercity rail network and allowing environmentally friendly travel, China should also try to optimize energy-saving of the transportation infrastructure and utilize energy-saving traffic technologies. Although the country has already achieved preliminary success, the energy saving of communication and transportation is still facing major issues. These include, underdeveloped transportation infrastructure; development of public transport is lagging behind; the lack of optimizing energy-saving management systems for transport; improvement required in traffic energy-saving policy systems; and the existing trend of consumers purchasing large-sized and high-fuel consuming SUV vehicles.

The transportation sector needs to study and develop a special action plan to establish a system of standards for energy-saving and emissions reduction. The sector needs to build standards for fuel consumption in shipping operations, port cargo handling machines and traffic construction machines and set a carbon emissions limit. It also needs to improve technical standards and implement energy-saving designs for highway bridges, engineering and greenery work. Legislation must be improved and energy-saving traffic management systems must be normalized and standardized. Older automobiles, locomotives and ships must be phased-out and automobile fuel quality improved. Finally, set a higher standard limit for fuel economy as found in other parts of the world to control the growth of heavy-emission cars.

A publication, "Research on medium and long-term energy-saving problems for China's transportation" [3], suggests that large energy-savings of up to 140 MTCE and 310 MTCE in 2020 and 2030, respectively, can potentially be achieved. Amongst this highway transportation and private cars will account for the majority of up to two-thirds. Research undertaken by the China Automotive Energy Research Center (CAERC), Tsinghua University, also drew a similar conclusion [4].

3. Results of Comprehensive Measurements and Calculations

Comprehensive measurements and calculations have a huge potential for generalized technology energy saving. The energy savings over fixed periods of each five-year plan, from 2010 to 2030, are shown in **Table 2**. The industrial sector has some potential in the first 15 years, which gradually reduces. The construction sector has huge energy-saving potential during the first 5 years, which later plateaus. The potential of the transportation sector gradually increases, will peak in 2015-2020, and then gradually drop. After 2010, the energy saving over fixed periods, in China's energy utilization sectors, increases from 200 MTCE to 250 MTCE, and then gradually reduces. Compared with 2010, the total potential energy saving will keep growing in 2015, 2020, 2025 and 2030 with values of 200, 450, 650 and 800 MTCE, respectively, albeit at a slow growth rate.

Table 2. Energy saving potential in China's energy utilization fields (x102MTCE/year).

	2010-2015	2015-2020	2020-2025	2025-2030	After 2030
Industrial sector	1	1	1	0.75	Gradually decreases
Construction sector	0.5	0.25	0.25	0.25	Plateaus
Transportation sector	0.5	1.25	0.75	0.5	Gradually decreases
Total	2.0	2.5	2.0	1.5	Gradually decreases
The year-end accumulative potential compared to 2010	2.0	4.5	6.5	8.0	Slowly increases

4. Policy Suggestions

Thus, an energy saving revolution would need integrations within the three sectors, *i.e.* industrial, construction and traffic. These include, the integration between key technologies and widely used technologies, between advanced technologies and flexible management, between energy-conservation technologies and comprehensive resource utilization technologies. In addition, promoting energy efficiency standards and introducing regulatory policies are also important tools.

Firstly, utilize advanced energy-saving technologies to gradually make China's industrial energy efficiency reach world-leading levels, and continue to refine mandatory national standards for energy consumption per unit product in key industries. Secondly, implement a green building plan to promote energy efficient building, and improve the level of energy efficiency with the support of regulations, technology, standards and design. Thirdly, promote energy-efficient transportation; speed up the construction of comprehensive transport systems; optimize transportation structure; encourage the use of energy efficient and environmentally friendly vehicles; promote the construction of energy efficient transportation infrastructure; increase investment in research and development of new energy vehicles; scientifically plan the construction of supporting facilities for gas refilling, battery charging, amongst other areas.

References

[1] National Bureau of Statistics of China (2013) China Statistics Yearbook 2013. China Statistics Press, Beijing.

[2] Building Energy Research Center of Tsinghua University (2013) Annual Report on Construction Energy-Saving Development in China. Building Energy Research Center of Tsinghua University, Beijing.

[3] Fu, Z.H., *et al.* (2011) Study on the Problems of Long-Term Energy Saving of Communication and Transportation of China. China Communication Press, Beijing.

[4] CAERC (2012) China Automotive Energy Outlook 2012. China Scientific Press, Beijing.

15

Life Cycle Energy of Low Rise Residential Buildings in Indian Context

Talakonukula Ramesh[1*], Ravi Prakash[2], Karunesh Kumar Shukla[3]

[1]Government Polytechnic, Nirmal, India
[2]Department of Mechanical Engineering, Motilal Nehru National Institute of Technology, Allahabad, India
[3]Department of Applied Mechanics, Motilal Nehru National Institute of Technology, Allahabad, India
Email: [*]rams2dg@gmail.com

Abstract

Life cycle energy of the building accounts for all energy inputs to the buildings during their intended service life. Buildings need to be constructed in such a way that energy consumption in their life cycle is minimal. Life Cycle Energy (LCE) consumption data of buildings is not available in public domain which is essentially required for building designers and policy makers to formulate strategies for reduction in LCE of buildings. The paper presents LCE of twenty (20) low rise residential buildings in Indian context. LCE of the studied buildings is varying from 160 - 380 kWh/m² year (Primary). Based on the LCE data of studied buildings, an equation is proposed to readily reckon LCE of a new building.

Keywords

Life Cycle Energy, Residential Buildings, Embodied Energy, Operating Energy

1. Introduction

Building construction sector is experiencing a fast-paced growth in developing countries, like India, due to growth of economy and rapid urbanization. A large number of buildings are being built for residential, commercial and office purposes every year. In India, 24% of primary energy and 30% of electrical energy is consumed in buildings [1]. The use of electricity in this sector is growing at the rate of 11% - 12% annually, which is 100% more than the average growth rate of 5% - 6% in the economy [2]. Besides the depletion of non-renewable energy sources, this energy use contributes greenhouse gases to the atmosphere, with consequent detrimental effects. In order to reduce the detrimental environment impacts of the buildings, new buildings need to be planned in such a way that energy consumption in their life cycle is minimal. In spite of the fast-paced growth of

---
[*]Corresponding author.

the building sector in India, Life Cycle Energy (LCE) consumption data for this sector is not available in the public domain; whereas a lot of work has been done in cold and western countries. Absence of macro-level data has been a barrier for the government to formulate effective policies to make the buildings energy-efficient.

Life cycle energy of the building accounts for all energy inputs to the buildings during their intended service life. It includes direct energy inputs during construction, operation and demolition phases of the buildings, and indirect energy inputs through the production of components and materials used in construction (embodied energy). If LCE is expressed in primary energy terms, it also gives a useful indication of environmental impacts attributable to buildings as primary energy consumption and associated emissions are proportional [3]-[5]. Life cycle energy evaluation of buildings becomes necessary not only for evaluating energy performance of the existing buildings but also to set a meaningful target for construction industry to construct new buildings with reduced energy demand, *i.e.* low energy buildings.

Low Energy Buildings

It is reported in different case studies available in the literature that operating energy of the buildings has largest share (80% - 90%) and embodied energy constitutes 10% - 20% in its life cycle energy distribution. Thus, the most important aspect for the design of buildings which demand less energy throughout their life cycle (low energy buildings) is the reduction in operating energy [6]-[8]. In order to reduce operational energy demand of the buildings, passive and active measures such as providing higher insulation on external walls and roof, using gas filled multiple pane windows with low emissivity coatings, ventilation air heat recovery from exhaust air, heat pumps coupled with air or ground/water heat sources, solar thermal collectors and building integrated solar photovoltaic modules, etc. can be used. But, reduction in operating energy is generally accompanied by increase in embodied energy of the buildings due to energy intensive materials used in the energy saving measures and on-site power generating equipment integrated with building.

Though embodied energy constitutes only 10% - 20% to life cycle energy, opportunity for its reduction should not be ignored. There is a potential for reducing embodied energy requirements through use of materials in the construction that requires less energy during manufacturing [9]. While using low energy materials, attention must be focused on their thermal properties and longevity as they have impact on energy use in operating phase of a building's life cycle. Thus, energy saving measures aimed at reducing one phase of energy use (operating) has impact on other phase of energy use (embodied energy) of the building. Hence, holistic evaluation of the buildings covering all phases of energy use is required to assess energy performance of the buildings. Another opportunity for reducing embodied energy is through use of recycled materials in the construction.

The present paper focuses on evaluation and presentation of LCE data of low rise residential buildings in Indian context. LCE of the buildings was evaluated for existing (conventional) and modified designs. Building designs are modified by applying energy saving measures viz. thermal insulation on wall and roof, double pane glass for windows and with on-site power generation equipment (PV modules). Such a study is expected to be useful for building designers and policy makers for holistic evaluation of buildings from life cycle perspective.

2. Methodology

A total of 20 house designs (**Table 1**) are obtained from house builders, consultants and owners of the buildings. All buildings are conventional houses with RCC frame work, walls filled with fired clay bricks, and RCC roof. The buildings are categorized by number of floors they have viz. one storey, two storey, and multi storey. Each floor contains one or more family portions consisting of bed rooms, drawing room, living room, and a kitchen. Bedrooms and living hall are air conditioned. The information of buildings such as usable floor area, conditioned area, number of families living, operating hours, etc have been collected.

Electricity from the national grid is being used for all operations of the buildings like running air conditioners, domestic appliances, water heating and lighting etc. The indoor operating set point temperature is around 25°C for cooling, 18°C for heating and all lighting controls of the building are manual. Bed rooms and living hall are air conditioned using window air conditioners having COP of 3 for cooling and 0.9 for heating (electrical resistance heating) for design conditions. Though, electrical resistance heating is not advisable, it is common in India, as harsh winter in most parts of the country lasts only for one or two months and people do not use heat pump or boiler for heating. The air conditioner utilization is about 11 hours on an average for bedrooms and 4 hours for the living room starting in the evening hours for all working days. On holidays, air conditioners start working in the afternoon 13.00 hours onwards. Detailed estimation of energy required for the production (embodied energy-

EBE) and operation phases of the buildings from a primary energy perspective is being considered. LCE of the buildings are evaluated for different locations (Allahabad, Ahmedabad, Hyderabad, Chennai and Bangalore) under different climatic zones of India viz: hot and dry, warm and humid, moderate, and composite (**Figure 1**).

Table 1. Details of the buildings studied.

BIN	Name	Category	Floor Area (m²)	Conditioned area (m²)	Description	Location
1	Resha	One storey	80	36	Single family, 3 BR house	Hyderabad
2	Harish	One storey	90	42	Single family, 2 BR house	Hyderabad
3	Janardhan	One storey	102	55	Single family, 2 BR house	Hyderabad
4	Goud	One storey	86	47	Single family, 2 BR house	Hyderabad
5	Eashwer	One storey	185	104	Two families, 2BR portion-1, 1BR portion-1	Hyderabad
6	Srinivas	One storey	155	102	Two families, single BR portions-2	Hyderabad
7	Ravindra	One storey	107	71	Single family, 2BR house	Hyderabad
8	Adil	One storey	62	46	Two families, single BR portions-2	Hyderabad
9	Keerthi	One storey	104	86	Single family, 3BR house	Hyderabad
10	Abhishek	Two storey	256	136	Two families, 3BR portions-2	Hyderabad
11	Alwal	Two storey	135	80	Two families, single BR portions-2	Hyderabad
12	Nirmal	Two storey	235	155	Two families, 3BR portions-2	Hyderabad
13	Mahipal	Two storey	268	180	Multy families, single BR flats-8	Hyderabad
14	Anand	Duplex	183	100	Single family, 4BR house	Hyderabad
15	RG	Duplex	175	120	Single family, 4BR house	Hyderabad
16	Mahendra	Duplex	450	340	Single family, 4BR house	Ahmedabad
17	Kiran Arcade	Multi storey	1286	600	Multy families, single BR flats-15	Hyderabad
18	Renuka	Multi storey	590	350	Multy families, two BR flats-12	Hyderabad
19	Pradeep	Multi storey	854	430	Multy families, single BR flats-12	Hyderabad
20	Rock town	Multi storey	1280	1024	Multy families, 3BR flats-4, 2BR flats-8	Hyderabad

BIN: Building Identification Number.

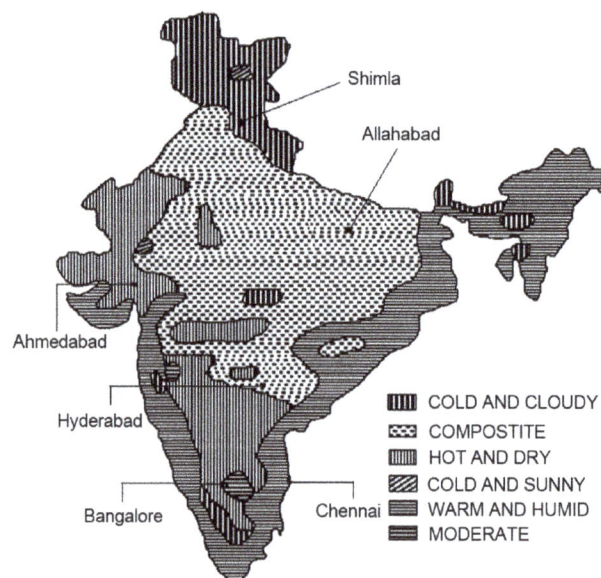

Figure 1. Map showing locations of the cities under different climatic zones of India.

Life Cycle Energy

LCE demand of the building is taken as the sum of the embodied energy of materials used in the construction (EBE) and operating energy (OPE) on an assumed lifespan of 75 years using following relation [10] [11]:

$$LCE = \sum m_i M_i + E_A L_b \tag{1}$$

where

m_i = Quantity of building material (i),

M_i = Embodied energy of material (i) per unit quantity (**Table 2**),

E_A = Annual Operating Energy (primary), L_b = Lifespan of the building (75 years).

Energy used for on-site construction and demolition at the end of its service life are ignored in the study as they contribute little (1%) to LCE [10]-[14]. Unit for LCE is chosen as kWh (thermal). However, normalized LCE per unit floor area and per year is useful for quick comparison of energy performance of buildings of different sizes or different design versions of a building. Hence, LCE and other energy entities (OPE and EBE) of the building are normalized to kWh/m^2 year based on their floor area and assumed lifespan of 75 years. Quantity of materials is estimated from the technical drawings of the buildings using QE-Pro software [15]. Embodied energy per unit quantity of building materials are compiled from literature [16]-[19].

The energy used for the renovation of buildings is included in EBE of the building. Annual electricity demand of the building is estimated by energy simulation of the building using dynamic energy simulation tool design builder [20]. The evaluated energy (electricity) demand of the buildings is then converted into primary energy using a conversion factor of 3.4 [21] for the Indian context and is termed as annual operating energy (E_A). Annual operating energy of the building is assumed to be same in future throughout its life span.

LCE demand is estimated for existing (conventional-Case A) and modified designs of the buildings for different climatic conditions of India. Building designs are modified by applying energy saving measures: adding 5 cm thick thermal insulation to wall and roof, and double pane glass for windows (Case B). LCE demand of the conventional building under particular climatic condition is taken as the base case for calculating energy savings. Further, LCE of the buildings is also evaluated with on-site power generating equipment (PV system). The embodied energy of PV modules, for initial installation and replacement, is included in calculation of EBE of the building. Number of times the PV modules are replaced is calculated using following relation:

$$N = (L_b / L_i - 1) \tag{2}$$

where

N = No of times the PV modules are replaced in life span of building,

L_b = Lifespan of the building,

L_i = Lifespan of PV modules (**Table 3**).

Table 2. Embodied energy of building materials.

Name of the Material	Unit	Embodied Energy per Unit (GJ)	Reference Source
Cement	ton	6.7	[16]
Steel	ton	28.212	[16]
Fired clay bricks	m^3	2.235	[16]
Aggregate	m^3	0.538	[16]
Glass	ton	25.800	[17]
Copper	ton	110.000	[18]
Ceramic tiles	ton	3.333	[16]
PVC	ton	158.000	[16]
Marble/Granite	ton	1.080	[19]
AC blocks	m^3	0.818	[16]
Fly ash bricks	m^3	1.341	[19]
Expanded polystyrene (EPS)	m^3	2.500	[19]
Aluminum	ton	236.8	[18]

Table 3. Particulars of PV modules.

Parameter	Value
Wattage per module	75 W_p
Short circuit current I_{sc}	4.8 A
Open circuit voltage V_{oc}	21 V
Maximum current I_{max}	4.5 A
Maximum voltage V_{max}	16.5 V
Area of single module	0.6 m^2
Type of cell	Single crystalline silicon
Number of cells in a module	36
Life span	30 years
Embodied energy of PV system (primary)	1710 kWh/m^2

Electricity generated from PV modules is simulated using e-Quest software [22] for different climatic conditions of India. PV modules and storage devices (batteries) are designed as explained in the reference [23]. Specifications and other particulars of PV modules are shown in **Table 3**.

3. Results and Discussion

The results obtained from the life cycle energy analysis of the buildings under different conditions are presented herein. **Table 4** presents the life cycle energy (LCE) demand of the conventional buildings studied under different geographical locations of India. LCE of the buildings is varying from about 160 - 380 kWh/m^2 year. There is wide variation in LCE demand of the buildings. The reasons for this variation could be attributed to differences in climatic conditions, conditioned floor area and layout of the buildings. However, LCE range of buildings for composite, hot and dry, warm and humid climates is almost same and it is about 200 - 380 kWh/m^2 year whereas for moderate climate it is 160 - 270 kWh/m^2 year. Single storey houses require higher LCE than two and multi-storey house under similar operating and climatic conditions. This is due to the fact that, single storey houses require higher operating energy than two and multi-storey houses as they have higher external surface area per m^2 of usable floor area which results in higher thermal load and energy consumption by cooling and heating equipment. Besides this, embodied energy of single storey houses is also higher than two and multi-storey houses. With increase in number of floors, external surface area per usable floor area comes down and hence multi-storey houses show better energy performance among the three.

Figures 2-4 show the variation of annual operating (electrical) energy demand of the buildings with conditioned floor area for different locations. It is observed that annual operating energy demand of the building is increasing with increase in conditioned floor area. Regression analysis is performed to obtain a relation between annual operating energy (electrical energy) and conditioned floor area of the buildings. A second order polynomial equation ($R^2 = 0.98$) can be best fit curve among the others-linear ($R^2 = 0.97$) and exponential ($R^2 = 0.8$). The relation between conditioned floor area and annual operating energy cannot be linear at higher conditioned floor areas. The reason is generally higher conditioned floor areas exist in multi-floor buildings; with increase in number of floors, external surface area per unit floor area of the building comes down thereby reducing the rate of increase in air conditioning load and corresponding operating energy of the building. Hence, the relation between conditioned floor area and annual operating energy becomes non linear with increase in conditioned floor area.

Hence, second order polynomial equation can be chosen to estimate annual operating primary energy (E_A) of the buildings.

$$E_A = 3.4\left(AX^2 + BX + C\right) \tag{3}$$

where
X = Conditioned floor area of the building (m^2),
A, B and C are regression coefficients and are shown in **Table 5**.

Further, it is observed that embodied energy of the buildings for single storey buildings is varying from 25 to 30 kWh/m^2 year (average 27.5 kWh/m^2 year) and for two and multi storey houses it is varying from 18 to 25 kWh/m^2 year (average 22 kWh/m^2 year) As variation in embodied energy of the buildings is not high, the average of the above values are taken as standard to represent embodied energy of single, two and multy-storey houses respectively.

Table 4. LCE demand of the residential buildings for different locations.

BIN	Name	Embodied energy kWh/m^2 year	Life cycle energy kWh/m^2 year				
			Hyderabad	Ahmedabad	Allahabad	Chennai	Bangalore
1	Resha	29.4	265	276	304	313	226
2	Harish	27.6	232	269	270	274	198
3	Janardhan	29	193	218	219	209	165
4	Goud	28	203	242	243	235	164
5	Eashwer	21	267	293	288	300	247
6	Srinivas	25	259	298	297	301	223
7	Ravindra	25.2	269	304	309	310	230
8	Adil	27.4	294	330	346	335	249
9	Keerthi	28	327	376	368	357	254
10	Abhishek	24.2	246	280	280	288	201
11	Alwal	18.5	266	297	291	290	197
12	Nirmal	23.5	271	305	315	300	230
13	Mahipal	18.3	278	318	325	322	225
14	Anand	21.5	255	285	288	294	207
15	RG Reddy	22	276	318	303	315	221
16	Mahendra	25	301	334	332	345	256
17	Kiran Arcade	22	247	272	276	280	210
18	Renuka	25	298	336	334	347	243
19	Pradeep	21	230	255	260	264	192
20	Rock town	23	317	349	346	364	269

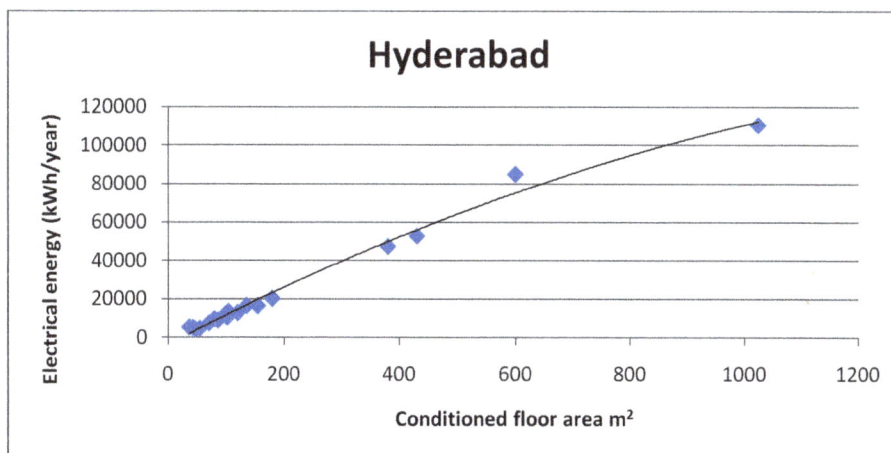

Figure 2. Variation of electrical energy demand of the buildings with conditioned floor area (Hyderabad location).

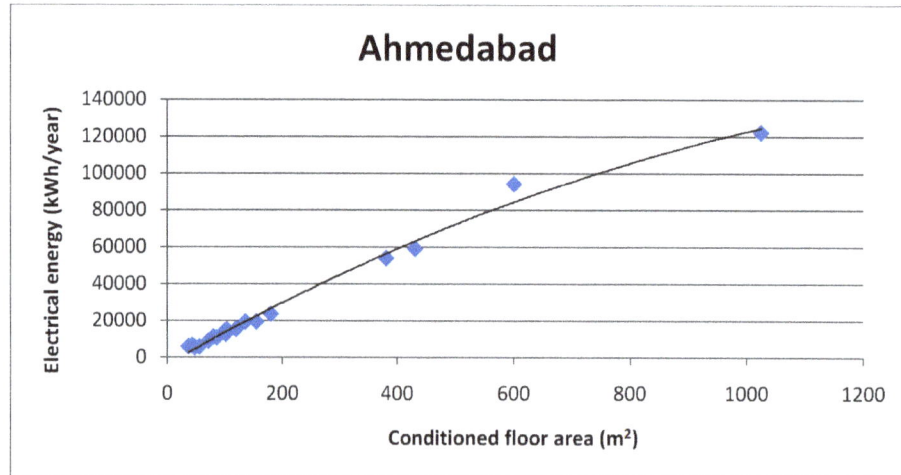

Figure 3. Variation of electrical energy demand of the buildings with conditioned floor area (Ahmedabad location).

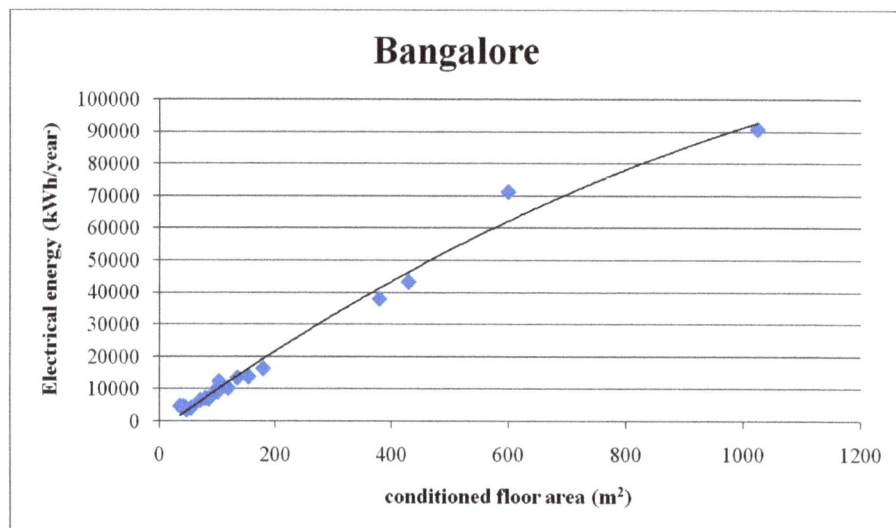

Figure 4. Variation of electrical energy demand of the buildings with conditioned floor area (Bangalore location).

Table 5. Regression coefficients for different locations.

	Hyderabad	Ahmedabad	Allahabad	Chennai	Bangalore
A	−0.043	−0.05	−0.055	−0.049	−0.035
B	157.4	177.3	182.1	182.4	129.6
C	−3388	−3601	−3951	−3941	−2925

Thus, to estimate LCE of the conventional buildings (in kWh/m² year) following equations is proposed:

$$LCE = EBE + E_A / FA_R \qquad (4)$$

where

EBE = 27.5 for single storey houses; 22 for two and multi-storey houses,

FA_R = Floor area (usable) of the building.

Tables 6-8 present LCE of the buildings with passive features (thermal insulation on envelope and double pane glass for windows) for different locations. LCE savings with passive features is about 5% - 30% depending

Table 6. LCE and savings % (values shown in parenthesis) from energy saving measures (Ahmedabad location).

BIN	Name	Case A	Case B
1	Resha	309	256 (17.2)
2	Harish	273	232 (15)
3	Janardhan	218	174 (20.2)
4	Goud	245	203 (17.1)
5	Eashwer	293	260 (11.3)
6	Srinivas	298	243 (18.5)
7	Ravindra	305	254 (16.7)
8	Adil	330	274 (17)
9	Keerthi	376	265 (29.5)
10	Abhishek	279	256 (8.2)
11	Alwal	297	240 (19.2)
12	Nirmal	304	273 (10.2)
13	Mahipal	318	282 (11.3)
14	Anand	285	257 (9.8)
15	RG Reddy	318	285 (10.4)
16	Mahendra	334	312 (6.6)
17	Kiran Arcade	271	261 (3.7)
18	Renuka	336	310 (7.7)
19	Pradeep	258	237 (8.1)
20	Rock Town	349	335 (4)

Table 7. LCE and savings % (values shown in parenthesis) from energy saving measures (Hyderabad location).

BIN	Name	Case A	Case B
1	Resha	265	231 (12.8)
2	Harish	235	212 (9.8)
3	Janardhan	193	163 (15.5)
4	Goud	203	173 (14.8)
5	Eashwer	267	249 (6.7)
6	Srinivas	259	226 (12.7)
7	Ravindra	269	237 (11.9)
8	Adil	294	253 (13.9)
9	Keerthi	327	242 (26)
10	Abhishek	246	237 (3.7)
11	Alwal	266	220 (17.3)
12	Nirmal	271	254 (6.3)
13	Mahipal	278	257 (7.6)
14	Anand	255	238 (6.7)
15	RG Reddy	276	260 (5.8)
16	Mahendra	301	289 (4)
17	Kiran Arcade	247	244 (1.2)
18	Renuka	298	287 (3.7)
19	Pradeep	230	219 (4.8)
20	Rock Town	317	312 (1.6)

BIN: Building Identification Number.

on the type, layout, and conditioned floor area of the buildings and also climatic conditions of locality. Single storey houses have better LCE savings than two and multi-storey houses because reduction in thermal load per unit floor area, due to thermal insulation on envelope, is higher for single storey houses than two and multi-storey houses.

Table 9 presents LCE savings of a single storey house with varying number of PV modules (on-site power generation) in combination with passive features. There is 30% to 70% reduction in LCE of the building. Use of PV modules seems to be most promising for primary energy reduction of the buildings.

Table 8. LCE and savings from energy saving measures for other locations.

Name of the building	Allahabad		Chennai		Bangalore	
	LCE	Savings %	LCE	Savings %	LCE	Savings %
Resha	251	17	259	17	209	8
Harish	233	14	236	14	187	6
Janardhan	175	20	169	19	152	8
Goud	200	18	191	19	155	5
Eashwer	259	10	260	13	233	6
Srinivas	243	18	238	21	211	5
Ravindra	256	17	264	15	218	5
Adil	282	18	271	19	229	8
Keerthi	263	29	251	30	206	19
Abhishek	252	10	265	8	195	3
Alwal	235	19	238	18	171	13
Nirmal	275	13	282	6	216	6
Mahipal	276	15	289	10	217	4
Anand	259	10	267	9	200	3
RG Reddy	283	7	293	7	216	2
Mahendra	305	8	316	8	241	6
Kiran Arcade	261	5	271	3	207	1
Renuka	310	7	324	7	240	1
Pradeep	237	9	245	7	188	2
Rock Town	331	4	354	3	266	1

Table 9. LCE and savings (values in bracket) for different cases of a house (Janardhan).

Case	Cities					
	Ahmedabad	Allahabad	Chennai	Bangalore	Hyderabad	Remarks
Case A	218	219	209	165	193	Conventional
Case B	174 (16)	175 (17)	169 (19)	152 (3)	163 (10)	Passive
Case B + 20 PV modules	128 (38)	131 (38)	120 (42)	111 (29)	117.4 (35)	On-site (part load)
Case B + 40 PV modules	72.4 (65)	76.8 (63)	65 (69)	62.4 (60)	62.5 (65)	On-site (part load)
Case B + Y No. PV modules	56.7 (73), Y = 60	56.7 (73), Y = 60	55 (74), Y = 52	55 (65), Y = 52	55 (70), Y = 52	On-site (self sufficient)

4. Conclusions

LCE of the buildings is varying from 160 - 380 kWh/m^2 year depending on the type (geometry) of the building and climatic conditions. With insulation on wall and roof along with double pane glass for windows, reduction in LCE of the buildings is about 5% - 30%. LCE of the buildings can be further reduced by on-site power generation from PV system (30 to 70%). A polynomial equation is proposed to readily reckon LCE of the new buildings. However, such equation needs to be improved when large number of LCE data is available in future.

The results of the present study are useful for building designers involved in design and construction of the energy efficient buildings and for policy makers to set meaningful targets. Some other cooling techniques like free cooling, evaporative cooling, solar air conditioning etc., may be tested to bring down LCE of the buildings. Use of energy efficient cooling/heating equipment and appliances would also reduce LCE of the buildings considerably.

References

[1] Bansal, N.K. (2007) Energy Security, Climate Change and Sustainable Development. In: Mathur, J., Wagner, H.J. and Bansal, N.K., Eds., *Science, Technology and Society: Energy Security for India*, Anamaya Publishers, New Delhi, 15-23.

[2] Ministry of Power, Government of India (2010) Performance Based Rating and Energy Performance Benchmarking for Commercial Buildings in India. http://www.powermin.nic.in/JSP_SERVLETS/internal.jsp

[3] Adalberth, K., Almgren, A. and Petersen, E.H. (2001) Life Cycle Assessment of Four Multi-Family Buildings. *International Journal of Low Energy and Sustainable Buildings*, **2**, 1-21.

[4] Kofoworola, O.F. and Gheewala, S.H. (2008) Environmental Life Cycle Assessment of a Commercial Office Building in Thailand. *International Journal of Life Cycle Assessment*, **13**, 498-511. http://dx.doi.org/10.1007/s11367-008-0012-1

[5] Junnila, S., Horvath, A. and Guggemos, A.A. (2006) Life-Cycle Assessment of Office Buildings in Europe and the United States. *Journal of Infrastructure Systems*, **12**, 10-17. http://dx.doi.org/10.1061/(ASCE)1076-0342(2006)12:1(10)

[6] Adalberth, K. (1999) Energy Use in Four Multi-Family Houses during Their Life Cycle. *International Journal of Low Energy and Sustainable Buildings*, **1**, 1-20.

[7] Winther, B.N. and Hestnes, A.G. (1999) Solar Versus Green: The Analysis of a Norwegian Row House. *Solar Energy*, **66**, 387-393. http://dx.doi.org/10.1016/S0038-092X(99)00037-7

[8] Citherlet, S. and Defaux, T. (2007) Energy and Environmental Comparison of Three Variants of a Family House during Its Whole Life Span. *Building and Environment*, **42**, 591-598. http://dx.doi.org/10.1016/j.buildenv.2005.09.025

[9] Yohanis, Y.G. and Norton, B. (2006) Including Embodied Energy Considerations at the Conceptual Stage of Building Design. *Power and Energy*, **220**, 271-288. http://dx.doi.org/10.1243/095765006X76009

[10] Fay, R., Treloar, G. and Iyer-Raniga, U. (2000) Life-Cycle Energy Analysis of Buildings: A Case Study. *Building Research & Information*, **28**, 31-41. http://dx.doi.org/10.1080/096132100369073

[11] Utama, A. and Gheewala, S.H. (2008) Life Cycle Energy of Single Landed Houses in Indonesia. *Energy and Buildings*, **40**, 1911-1916. http://dx.doi.org/10.1016/j.enbuild.2008.04.017

[12] Adalberth, K. (1997) Energy Use during the Life Cycle of Buildings: A Method. *Building and Environment*, **32**, 317-320. http://dx.doi.org/10.1016/s0360-1323(96)00068-6

[13] Adalberth, K. (1997) Energy Use during the Life Cycle of Single-Unit Dwellings: Examples. *Building and Environment*, **32**, 321-329. http://dx.doi.org/10.1016/s0360-1323(96)00069-8

[14] Treloar, G., Fay, R., Love, P.E.D. and Iyer-Raniga, U. (2000) Analysing the Life-Cycle Energy of an Australian Residential Building and Its Householders. *Building Research & Information*, **28**, 184-195. http://dx.doi.org/10.1080/096132100368957

[15] QE-Pro (2010) Quantity Estimation & Project Management Software. www.softtech-engr.com

[16] DA (Development Alternatives) (1995) Energy Directory of Building Materials. Building Materials & Technology Promotion Council, New Delhi.

[17] Reddy, B.V.V. and Jagadish, K.S. (2003) Embodied Energy of Common and Alternative Building Materials and Technologies. *Energy and Buildings*, **35**, 129-137. http://dx.doi.org/10.1016/s0378-7788(01)00141-4

[18] TERI (The Energy and Resources Institute) (2004) Sustainable Building Design Manual, Volume 2, Sustainable Building Design Practices. New Delhi, 91-112.

[19] Gupta, T.N. (1998) Building Materials in India. Building Materials & Technology Promotion Council, New Delhi.

[20] DesignBuilder-Building Design, Simulation and Visualization (2010) www.designbuilder.co.uk.

[21] TERI Press (2007) TEDDY TERI Energy Data Directory and Year Book 2005-06. New Delhi.

[22] E-Quest (2009) The Quick Energy Simulation Tool. http://www.doe2.com/equest

[23] Arvind, Ch., Tiwari, G.N. and Chandra, A. (2009) Simplified Method of Sizing and Life Cycle Cost Assessment of Building Integrated Photovoltaic System. *Energy and Buildings*, **41**, 1172-1180. http://dx.doi.org/10.1016/j.enbuild.2009.06.004

16

Method for Determination of Optimal Installed Capacity of Renewable Sources of Energy by the Criterion of Minimum Losses of Active Power in Distribution System

P. D. Lezhniuk[1], V. A. Komar[1], D. S. Sobchuk[2]

[1]Department of Electric Power Stations and Systems, Vinnytsia National Technical University, Vinnytsia, Ukraine
[2]Department of Electric Energy Supply, Lutsk National Technical University, Lutsk, Ukraine
Email: kvo76@mail.ru

Abstract

New method for determination of optimal placement and value of installed capacity of renewable source of energy (RES) by the criterion of minimum losses of active power, that allows taking into consideration the dependence of RES on natural conditions of region, schedule of energy supply, parameters and configuration of distribution network is suggested in the paper. Results of computations of test scheme confirm the efficiency of the proposed method and its simplicity as compared with the methods considered in literature sources.

Keywords

Distributed Generation; Renewable Sources of Energy; Minimum Losses of Active Power

1. Introduction

World electric power generation was traditionally developing by the means of centralization of generation system, construction of more powerful power plants and their integration in power complexes. As a result, large geographically extensive power systems were created: the European ENTSO-E, unified power system in Russia, unified power system in Ukraine and others. In the last few years, steady progress in energy sector of national economy restructuring greatly changed general concept of power branch development, it concerns the introduction of new ideology, namely introduction of distributed power generation. Distribution Generation System

(DGS) is defined as sources of electrical energy, connected directly with distribution system or connected to such network on the side of consumers. Wide spread of DG is connected with the introduction of high-efficient gas-turbine and steam-turbine plants [1] and development of renewable sources of energy. Among them the most widespread are wind electric plants (WEP) and photovoltaic systems (PVS).

Introduction of DG in electric networks, especially constructed on the base of renewable sources of energy (RSE), besides reduction of ecological impact l on the environment and solution of many problems connected with harmful emissions as a result of energy generation production, allows, first, sufficiently increasing the efficiency of resources utilization, and in future reducing the cost of electric energy, unload transmission lines and distribution systems.

Installation of DG sources in distributive electric networks near consumption centers changes direction of power flows. It is necessary to distinguish three situations concerning load nodes and DG [2]:

Power of each node in electric network is greater or equals the power of DG sources, connected to these nodes.

2. In electric network there exists one node, where the power of DG is greater than the power of this node load, but total power of DG sources of this network on the whole is less, than its total load.

3. In electric network there exists minimum one node, where DG power exceeds the power of this node load and total power of DG sources of this network on the whole is greater than its total load.

In the first case DG sources placed in electric network will influence the reduction of power losses in distribution networks. In the second case DG sources can permanently increase power losses in some lines of distribution network, but generally, total power losses are decreased. In the third case, total power losses of the distribution network will be higher than the prior installation of DG sources. Also different DG sources operate with different $\cos\varphi$ and their output reactive power can vary from insignificant generation (steam-turbine plants) to significant consumption (FES with asynchronous generators), that also negatively influences the magnitude of power losses in electric networks [3].

2. Analysis of the Research

Intensive introduction of distributed generation (DG) created a number of problems, one of which is the selection of the location for the connection to the network and their installed capacity. The solution of this problem enables to obtain necessary effect as a result of the introduction of dispersed generation source–reduction of power losses and enhancement of power quality.

The analysis of the existing approaches allows making a conclusion regarding the directions of the research. The existing methods are based on various approaches: analytical methods [4] [5]; mixed integer programming [6]; heuristic approach [7]; genetic algorithms [8], loss sensitivity factor method [9]. And it is only a part of research, performed in this direction. Each of the methods has its advantages, enabling to take into account certain peculiarities of distributed generation (DG) and renewable energy sources (RES). But none of the methods allows taking into account the dependence of renewable source of energy generation schedule on natural conditions. This factor is very important, especially for solving the problem of the selection of the installed capacity.

3. Characteristics of Renewable Sources of Energy

The comparison of the schedules of power consumption and generation, by such renewable sources of energy as photovoltaic (PV) and wind-electric plants (WEP) allows speaking about their low stability regarding the support of power balance (see **Figures 1** and **2**). In the problem of power balance support the capacities of PV and WEP can be referred to conventionally controlled sources. That is, theoretically it is possible to change generation, within the limits, dependent on natural conditions but, in this case their efficiency will be considerably reduced. That is why, it is expedient to develop such a method that would allow taking into consideration such characteristic feature of renewable sources of energy (RSE) in the problem of selection of their installed capacity.

To take into account the random component of generation schedules relatively the schedules of power consumption is possible by means of the analysis of the most probable generation values and the most probable values of consumption at the definite periods of the day. For each hour of the day the following dependences are constructed:

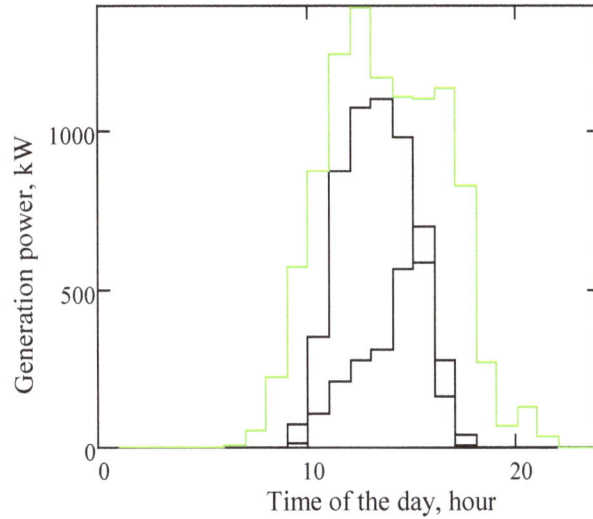

Figure 1. Seasonal change of daily schedule of PVS operation.

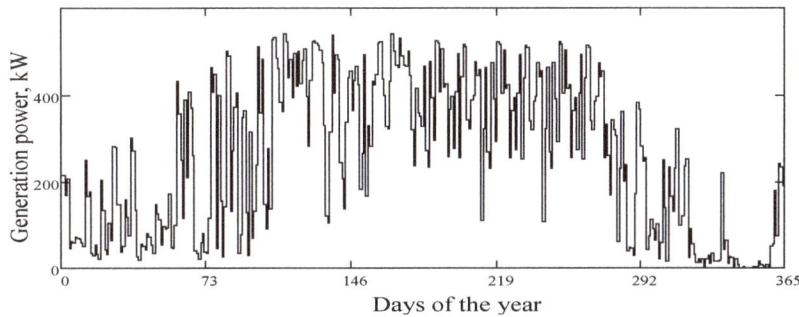

Figure 2. Change character of average values of PVP generation, determined by daily schedules during the year.

Depending on the number of corresponding levels occurrences (generation, consumption) their probabilities can be determined:

$$p = \frac{n}{365},$$

(1)

where n –is the number of levels occurrences (generation, consumption)during a year.

Having all the probabilities, the stability factor can be evaluated; this factor will characterize the probability of electric energy supply, generated by renewable source of energy, of the consumers, connected to corresponding feeder of distribution system. Instability factor can be determined by the expression:

$$k_{stab.} = \sum_{i=1}^{24} \left[p_{daysi} \sum_{j \in M} \left(p_{RES\_yr_j} \sum_{l \in N} p_{cons\_yr_l} \right) \right],$$

(2)

where p_{days} —is the probability of occurrence of the daily schedule stage $\left(p_{days} = \frac{1}{24} \right)$; p_{RES_yr} —is the probability of the occurrence of the generation stage during a year, is determined by the expression (1); M —is the set of non-zero stages; p_{cons_yr} —is the probability of the occurrence of the consumption stage during a year, is determined by the expression (1); N —is the set of consumption stages which are below the generation level of the corresponding period of the day (see **Figure 3**).

Stability factor enables to evaluate the possibilities of the source to cover the necessary consumption of electric energy. It depends on the power of electric energy source. For the accounting of energy supply it is suggested to introduce the coefficient, that is determined by the ratio of the expectation of annual consumption $M\left(W_{cons}\right)$ to expectation of RES annual generation $M\left(W_{RES}\right)$:

$$k_{pr} = \frac{M\left(W_{cons}\right)}{M\left(W_{RES}\right)}. \qquad (3)$$

Necessary expectations can be determined analyzing annual graphs by the duration (see **Figure 4**).

The suggested coefficients allow characterizing the source of electric energy relatively natural conditions and schedule of the consumption to be covered.

4. Method of Determination of RES Installed Capacity

The introduction of dispersed generation sources in the distribution systems has created a number of problems, namely, they must be considered not as main-radial but as networks with double sided supply or local electrical system (LES). That is why, to evaluate the necessary installed capacity of DG it is expedient to carry out the analysis of power flows distribution in LES.

For this purpose we consider the problem of determination of currents that provide minimum losses of active power in LES. In general case, if technical constraints with the current of generating nodes are missing, the objective function can be formulated in the following way:

Minimize

$$\Delta P = \hat{I}^{\mathrm{T}} R \hat{I} \qquad (4)$$

if

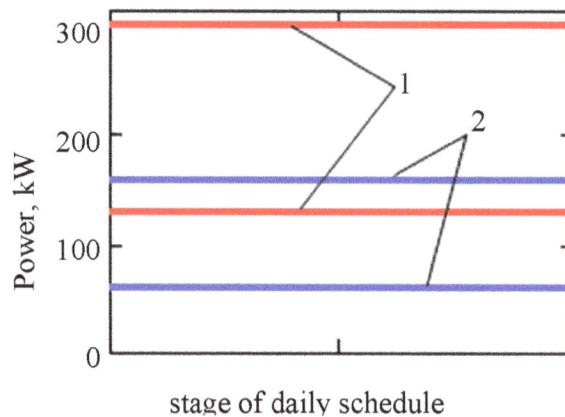

Figure 3. Probability levels of generation 1 and consumption 2 of the definite period of time during a year.

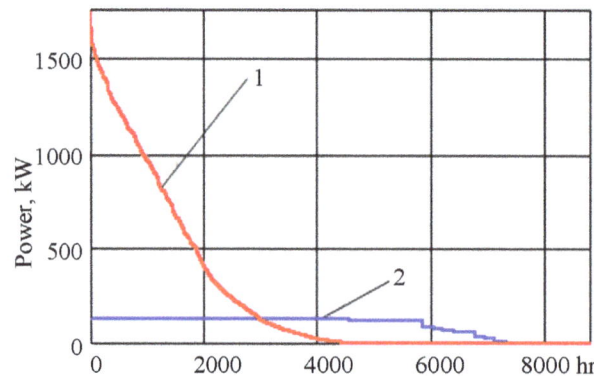

Figure 4. Annual schedules of generation 1 and consumption by duration.

$$\left.\begin{array}{l} M'I_a = J_a \\ M'I_q = J_q \end{array}\right\},$$
(5)

where $\dot{I}^T$, $\hat{I}$ —are transposed and conjugate vectors of currents in the branches; I_a, I_q—are vectors of active and reactive components of currents in branches; J_a, J_q—are vectors of active and reactive components of nodal currents; R—is diagonal matrix of resistances of the branches; M'—is the first matrix of the network connections, where lines corresponding to generating nodes are deleted (this is equivalent to the integration of all supply sources in one balancing node).

It should be noted, that the solution of the formulated optimization problem (4) on condition (5) allows determining not only optimal currents in the branches of the network but also establish optimal, from a viewpoint of active power losses minimization, supply sources loads If active, reactive or full load of any supply source is fixed, then in matrix M' the row, corresponding to this node, must be present and the current must be included in the vector of nodal currents with corresponding sign.

For determining minimum losses of active power and values of optimal currents in corresponding branches, we will make use of Lagrangian multiplier method. Lagrangian function for (4) taking into consideration the constraint Equations (5)

$$W = \Delta P + \left[\mu_a^T \mu_q^T\right] \begin{bmatrix} M'I_a - J_a \\ M'I_q - J_q \end{bmatrix},$$

where $\left[\mu_a^T \mu_q^T\right]$—is the transposed vector of Lagrangian undertemined multipliers.

From the condition of zero equality of partial derivatives W by variables, being optimized and Lagrangian multipliers we obtain the following system of Equations:

$$\begin{bmatrix} 2R & 0 & M'^T & 0 \\ 0 & 2R & 0 & M'^T \\ M' & 0 & 0 & 0 \\ 0 & M' & 0 & 0 \end{bmatrix} \begin{bmatrix} I_{a0} \\ I_{q0} \\ \mu_a \\ \mu_q \end{bmatrix} = \begin{bmatrix} 0 \\ 0 \\ J_a \\ J_q \end{bmatrix},$$

where optimal currents in the branches and Lagrangian multipliers:

$$\begin{bmatrix} I_{a0} \\ I_{q0} \\ \mu_a \\ \mu_q \end{bmatrix} = \left(\begin{bmatrix} 2R & 0 & M'^T & 0 \\ 0 & 2R & 0 & M'^T \\ M' & 0 & 0 & 0 \\ 0 & M' & 0 & 0 \end{bmatrix}\right)^{-1} \begin{bmatrix} 0 \\ 0 \\ J_a \\ J_q \end{bmatrix}.$$

where symbol -1 and T denote, correspondingly, inversion and transposition of the matrix.

Having divided the expression in brackets into blocks, as it is shown with heavy lines and applied Frobenius formula:

$$\begin{bmatrix} A & B \\ C & D \end{bmatrix}^{-1} = \begin{bmatrix} A^{-1} + A^{-1}BH^{-1}CA^{-1} & -A^{-1}BH^{-1} \\ -H^{-1}CA^{-1} & H^{-1} \end{bmatrix}$$

where $\begin{bmatrix} A & B \\ C & D \end{bmatrix}$—is $(m_1 + m_2) \times (m_1 + m_2)$ matrix, A—is $m_1 \times m_1$ nonsingular square matrix, D—is $m_2 \times m_2$ and $H = D - CA^{-1}B$ square matrix.

After simple transformations we obtain the solution of the problem of optimal currents determination in the branches:

$$\begin{bmatrix} I_{a0} \\ I_{q0} \\ \mu_a \\ \mu_q \end{bmatrix} = \begin{bmatrix} C_r & 0 \\ 0 & C_r \\ -2R_{ij} & 0 \\ 0 & -2R_{ij} \end{bmatrix} \begin{bmatrix} J_a \\ J_q \end{bmatrix},$$
(6)

where $C_r = R^{-1}M'^{\mathrm{T}}\left(M'R^{-1}M'^{\mathrm{T}}\right)^{-1}$ —is the matrix of current distribution coefficients of LES calculation scheme, where branches resistances are presented by their active components (r-circuit of LES); $R_{ij} = \left(M'R^{-1}M'^{\mathrm{T}}\right)^{-1}$ —is the matrix of nodal resistances of r-circuit of LES.

Vectors of active and reactive components of generating nodes optimal currents

$$J_{ao}^r = M''I_{ao};$$

$$J_{qo}^r = M''I_{qo},$$

where M''—is the matrix, lines of which are lines of the connections matrix, corresponding to generating nodes.

Proceeding from the solution of the problem (4), we can make a conclusion, that minimum losses of active power in LES for the case, when constrains are not imposed on nodal currents, occur when active and reactive components of currents are distributed in it depending on resistances, *i.e.*, according to r-equivalent circuit of LES. This result can be extended to power transfers.

Using the assumption that there are no transfers of reactive power and reactive components of equivalent circuit of network elements, we obtain the dependence in matrix form for active power transfers in the branches of the network:

$$P_B = -\frac{U_{nom}^2}{2}R^{-1}M^{\mathrm{T}}\mu_a; \qquad (7)$$

where R—is diagonal matrix of the resistances of the branches; M—is the first matrix of the connections; U_{nom}—is nominal voltage of the network.

From the system of Equations, obtained by differentiation with variables from (7), Lagrangian undetermined multipliers can be determined in the form:

$$\mu_a = -\frac{2}{U_{nom}^2}G_y^{-1}P; \qquad (8)$$

where $G_y = MR^{-1}M^{\mathrm{T}}$—is the matrix of nodal active conductances; P—is the vector of active powers in the nodes of the circuit.

After substitution of (8) into (7) we obtain

$$P_B = R^{-1}M^{\mathrm{T}}G_y^{-1}P. \qquad (9)$$

In (9) the expression $R^{-1}M^{\mathrm{T}}G_y^{-1}$ or $C_r = R^{-1}M^{\mathrm{T}}\left(MR^{-1}M^{\mathrm{T}}\right)^{-1}$ in correspondence with the above-mentioned, are optimal current distribution coefficients.

Rewrite (9) in the form

$$P_B = C_r^{\mathrm{T}}P. \qquad (10)$$

Similar results can be obtained taking into consideration the transfers of reactive power, that is why the Equation (10) will be valid for full power.

Using (10) we can develop the method of RES installed capacity determination by the criterion of minimum losses of active power.

5. Algorithm of the Method

The information to be used for the start of the computation is statistical data (minimum for the previous year) and the forecast, regarding natural conditions (solar radiation, wind flows); load curves of feeder's nodes, equivalent circuit and parameters of feeder's elements (see **Figure 5**).

6. The Results of Research

For validation of the performance of the developed method the feeder was considered, its circuit is shown in **Figure 6**.

Let us consider, as possible by technical conditions, nodes for RES 7, 10 and 20 connection. For the selection of the best, by energy losses criterion, we will determine current distribution coefficients for each of the variants.

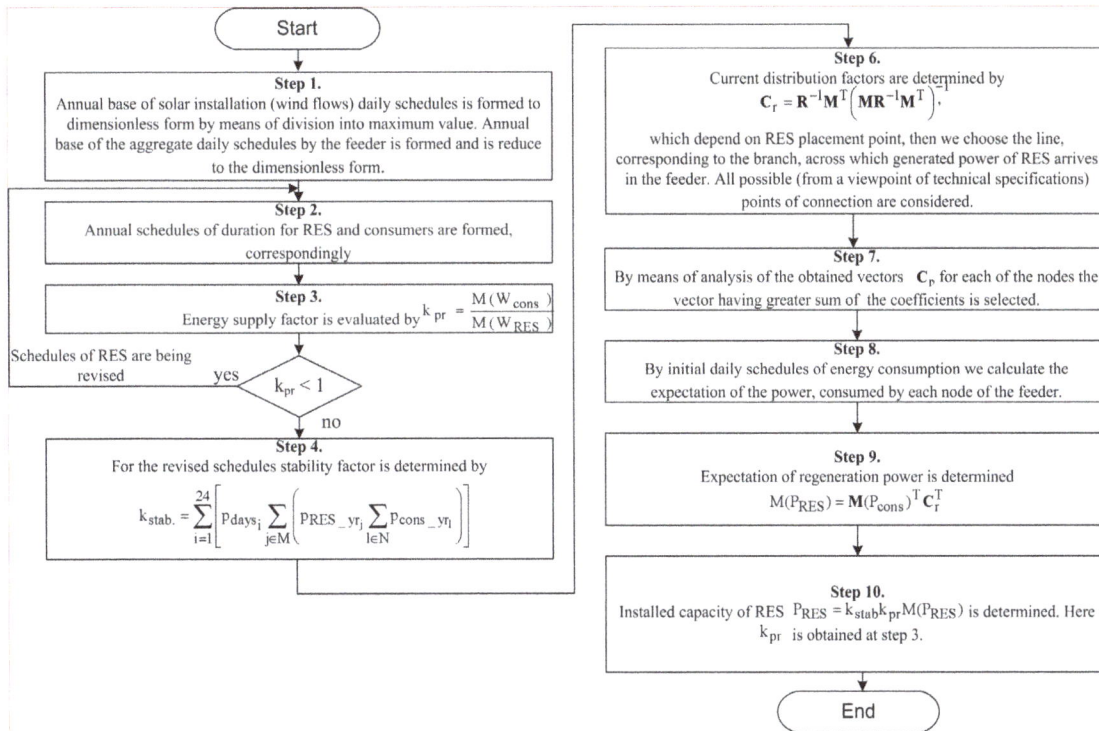

Figure 5. Algorithm of the method.

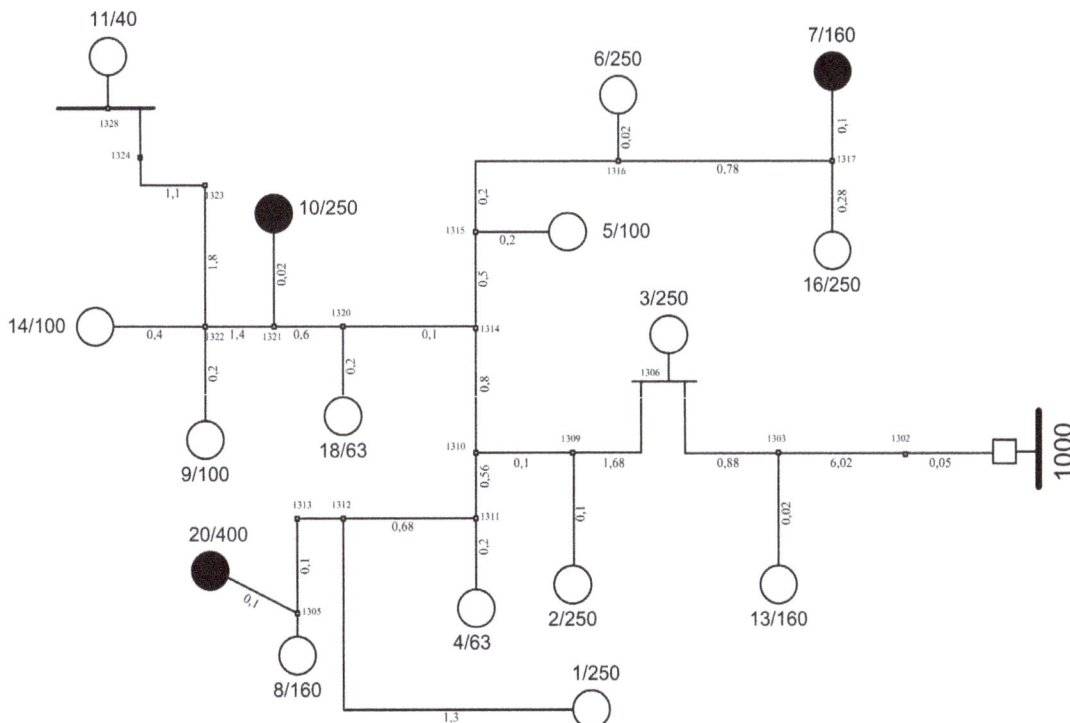

Figure 6. Fragment of distribution network.

Values of C_r coefficients are shown in **Table 1**. As the total value of the coefficients is greater for the node 10, then this node is the best node for RES connection. This is proved by the results of computations. From **Figure 7** (for node 20) and **Figure 8** (for node 10) the conclusion can be made regarding greater relief effect if

Table 1. Analysis of the vectors of current distribution coefficients.

	number	1	2	3	4	5	6	7	8	9	10	11	13	14	16	18	20	Σ
Values of current distribution coefficients	number relatively to node 20	0.944	0.568	0.242	0.753	0.596	0.596	0.596	0.972	0.596	0.596	0.596	0.071	0.596	0.596	0.596	1	9.914
	number relatively to node 10	0.661	0.63	0.269	0.661	0.834	0.834	0.834	0.661	0.994	1	0.994	0.079	0.994	0.834	0.865	0.661	11.81
	number relatively to node 7	0.466	0.444	0.189	0.466	0.763	0.807	1	0.466	0.588	0.588	0.588	0.056	0.588	0.978	0.588	0.466	9,042

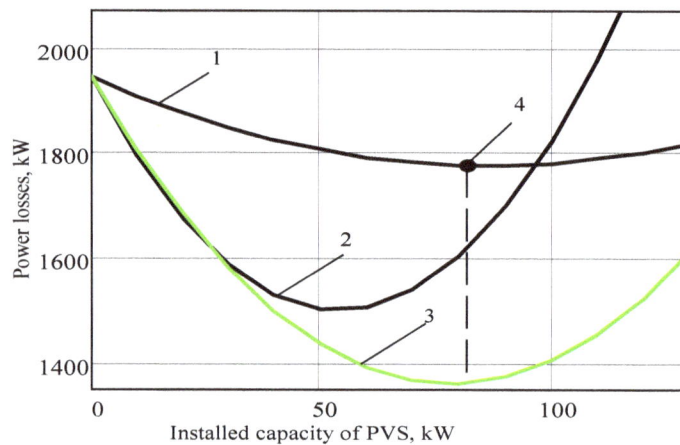

Figure 7. Computation results for the case of PVP placement in node 20.

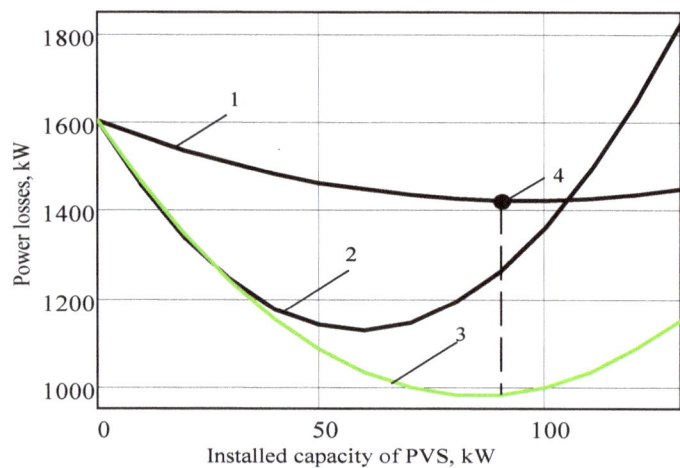

Figure 8. Computation results for the case of PVS placement in node 10.

PVS is installed in node 10 than in node 20.

For comparison the computation of active power losses during the year was carried out by means of exhaustive search of different powers of PVP for the investigated schedule of generation and consumption.

Graphic interpretation of the results is shown in **Figure 7** (for node 20) and **Figure 8** (for node 10)

In **Figures 7-9** curve 1—change of active power losses during a year, obtained by means of exhaustive search

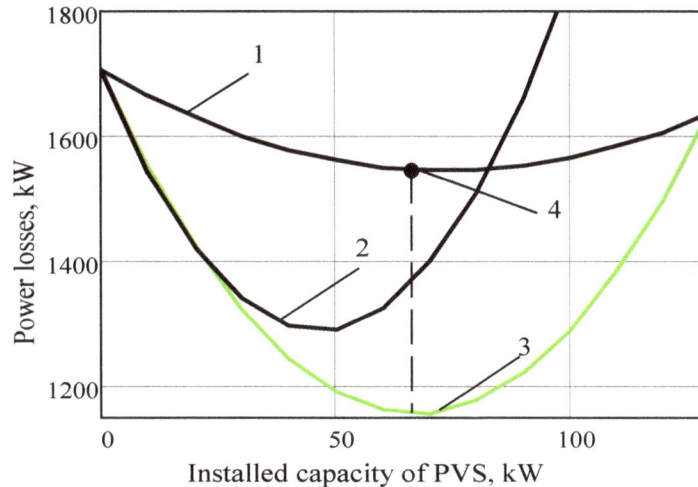

Figure 9. Computation results for the case of PVS placement in node 7.

of the installed capacities of PVS; curve 2—change of active power loss during a year, obtained as a result of exhaustive search of installed capacities s of PVS on condition that the installed capacity of the source is selected as a result of analysis of the most probable stage of power consumption; curve 3—change of active power loss during a year, obtained as a result of exhaustive search of the installed capacities of PVS on condition, that the operation mode of the plant is stipulated not only by natural conditions but also by the schedule of power consumption.

7. Conclusions

The suggested method of determination of optimal installed capacity of RES allows taking into consideration the dependence of their operation schedule on natural conditions, point of connection to distribution grid and loading curve. The algorithm of the method is easily formalized and realized in the form of programming product, which distinguishes it from the methods considered in the survey.

Calculations performed, prove the validity of the method both regarding the problem of determination of the point of RES connection and regarding the determination of the optimal power of the source by the criterion of minimum losses of electric energy.

References

[1] Bieliaiev, L.S., Lahariev, A.V. and Posiekalin, V.V. (2004) Power Engineering of XXI Century: Conditions of Development, Technologies, Forecasts. Science, Novosibirsk, 386.

[2] Ackerman, T., Knyazkin, V. (2000) Interaction between Distributed Generation and the Distribution Network. *Transmission and Distribution Conference and Exhibition: Asia Pacific IEEE/PES*, 1357-1362.

[3] Gonzalex-Longatt, F. (2007) Impact of Distributed Generation over Power Losses on Distribution System. *9th International EPQU Conference*, Barcelona.

[4] Acharya, N., Mahat, P. and Mithulananthan, N. (2006) An Analytical Approach for DG Allocation in Primary Distribution Network. *International Journal of Electrical Power & Energy Systems*, **28**, 669-678.
http://dx.doi.org/10.1016/j.ijepes.2006.02.013

[5] Hung, D.Q., Mithulananthan, N. and Bansal, R.C. (2010) Analytical Expressions for DG Allocation in Primary Distribution Networks. *IEEE Transactions on Energy Conversion*, **25**, 814-820.
http://dx.doi.org/10.1109/TEC.2010.2044414

[6] Chang, R.W., Mithulananthan, N. and Saha, T.K. (2011) Novel Mixed-Integer Method to Optimize Distributed Generation Mix in Primary Distribution Systems. 2011 21st Australasian Universities Power Engineering Conference (AUPEC), Brisbane, 25-28 September 2011, 1-6.

[7] Abu-Mouti, F.S. and El-Hawary, M.E. (2011) Heuristic Curve-Fitted Technique for Distributed Generation Optimisation in Radial Distribution Feeder Systems. *IET Generation, Transmission & Distribution*, **5**, 172-180.

http://dx.doi.org/10.1049/iet-gtd.2009.0739

[8]   Singh, D. and Verma, K.S. (2009) Multiobjective Optimization for DG Planning with Load Models. *IEEE Transactions on Power Systems*, **24**, 427-436. http://dx.doi.org/10.1109/TPWRS.2008.2009483

[9]   Griffin, T., *et al.* (2000) Placement of Dispersed Generation Systems for Reduced Losses. *Proceedings of the 33rd Annual Hawaii International Conference on System Sciences*, Hawaii, 4-7 January 2000.

17

Energy Scenario: Production, Consumption and Prospect of Renewable Energy in Australia

A. K. Azad*, M. G. Rasul, M. M. K. Khan, T. Ahasan, S. F. Ahmed

School of Engineering and Technology, Central Queensland University, Rockhampton, Australia
Email: *a.k.azad@cqu.edu.au

Abstract

Australia is the world's 9th largest energy producer, 17th largest consumer of non-renewable energy resources and ranks 18th on a per person energy consumption basis. Australia's energy consumption is primarily composed of non-renewable energy resources (coal, oil, gas and related products), which represent 96% of total energy consumption. Renewables, the majority of which is bioenergy (wood and wood waste, biomass, and biogas) combined with clear energy namely wind, solar hot water, solar electricity, hydroelectricity account for the remaining 4% consumption. Australia's renewable energy resources are largely undeveloped which can contribute directly to the Australian economy. In this article, a review of literature on energy scenario is presented and discussed. Australia's total energy production, consumption, storage and export (including renewable and non-renewable) data has been analyzed and discussed in this study. The main objective of the study is to analyze the prospect of renewable energy in Australia. This study concludes that Australian economy will grow faster if its undeveloped renewable energies can be used efficiently for electricity generation and transport sector.

Keywords

Renewable Energy; Non-Renewable Energy; Total Energy Production; Electricity Generation; Economy

1. Introduction

The world is facing two major problems, namely energy and environment. Another issue which is closely linked with the above is economy. So, energy, economy and environment are bonded by three dimensional relationships with bi-directional causal relationship among them [1]. The world energy consumption is likely to grow faster than the increase in the population [2]. The International Energy Outlook-2013 projects that world energy consumption will be grown by 56% between 2010 and 2040. Total world energy use will rise from 524 quadrillion British thermal units (Btu) in 2010 to 630 quadrillion Btu in 2020 and to 820 quadrillion Btu in 2040 [3].

---

*Corresponding author.

Much of the growth in energy consumption occurs in countries outside the Organization for Economic Cooperation and Development (OECD), known as non-OECD, where demand is driven by strong, long-term economic growth. Energy use in non-OECD countries increases by 90%, in OECD countries, the increase is 17%. The International Energy Outlook 2013 reference case does not incorporate prospective legislation or policies that might affect energy markets [3]. In the world energy status, Australia is holds 9th position in energy production and 17th position as an energy consumer [4,5]. The total energy production including energy export was 17,460 Petajoulesin 2011-2012 which is equivalent to total energy production increase by 5% relative to 2010-2011 [6].

Research is ongoing throughout the world on how to fulfil the energy demand successfully without hampering environment of our planet. Many of them deal with the issue of energy and environment, including the four main measures, namely energy saving and efficiency; switching to natural gas; CO_2 recovery; development of alternative energy sources to reduce the CO_2 emissions resulting from energy use [7]. Energy generation from the combustion of fossil fuels has simultaneously created several environmental concerns which can threaten the sustainability of our ecosystem. One of the primary concerns is the emissions of greenhouse gases and other types of air pollutants such as hydrocarbons, nitrogen oxide and volatile organic compounds [8]. To minimize the greenhouse effect the world is moving towards alternative energy sources which are eco-friendly, clear, and green energy [9-11].

The population growth is one of the major causes for high energy demand. The United Nations (UN) predicts the world population to reach 9 billion by 2030 [12]. Growth in population and the ever-increasing development of new production technology, new transport, living standard, industrialization etc. is leading to rising energy use [13]. However, development of a country is presently indexing by their energy consumption. But the fossil fuel which is the major sources of energy is decreasing gradually and the harmful effect is increasing sharply day-by-day [14,15]. Australia is presently consuming 96% of non-renewable energy and renewables for the remaining 4%. Australia has 33% of the world's uranium resources, 10% of world black coal resources and almost 2% of world conventional gas resources. It has only a small proportion of world crude oil resources. But there is also potential for a number of emerging clear energy technologies that are yet to be commercially developed, including large scale solar energy plant, geothermal generation technologies, ocean energy technologies and carbon capture and storage to reduce emissions from coal, oil and gas. Modelling by the Australian Energy Market Operator shows that 100% of power from clean energy would be technically viable by 2030-although with a price tag ranging from $219 billion to $252 billion. Australian Energy Market Operator (AEMO) investigated two future scenarios featuring an electricity grid fuelled entirely by renewable resources in 2030 and 2050. This is the first study of its kind by AEMO into potential costs and the viability of moving to an electricity generation system fuelled entirely by renewable resources [16]. The study aimed to investigate the present energy scenario in Australia. The total energy production by fuel type, energy consumption by fuel type, energy export and prospect of renewable energy data has been analysed and discussed in next sections.

2. Australian Energy Scenario

2.1. Primary Energy Production by Fuel Type

Figure 1 shows the total primary energy production in Australia. In 2011-12 total energy production (17,460 pJ) increased by 5% with respect to 2010-11 (16640 pJ) including exported energy. It has about 37% of domestic consumption and 63% of net energy export [6]. So, Australia's net energy production serves both domestic and international market. Energy demand increased gradually in both markets has spurred strong growth by 9% per year between 2000-01 and 2011-12 [4]. Presently, more than 60% of primary energy accorded from coal. So, Australian energy economy is fully dependent on coal energy which is clearly shown in **Figure 2**.

2.2.Primary Energy Consumption by Fuel Type

Total energy consumption is calculated as original production plus import less export and change in stock. So, in a word, the total energy used within the Australian economy is called net energy consumption. In 2011-12, total energy consumption increased by 2% with respect to 2010-11 and rises to 6193 pJ. Significant growth of energy consumption was found in petroleum sectors. It was contributed 39% of total energy consumption in 2011-12 [6]. The lowest relative consumption of black and brown coal together accounted 34% of total energy consumption since early 1970s. At a glance the energy growth by fuel type is shown in **Table 1**.

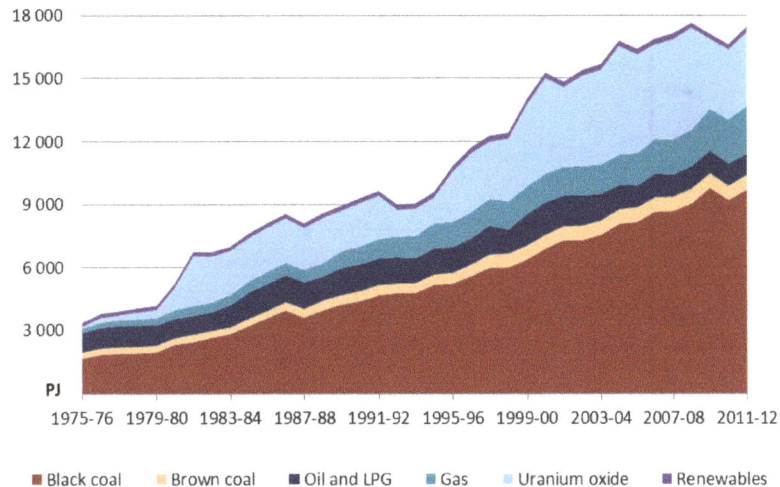

Figure 1. Total primary energy production in Australia.Source: Energy in Australia 2013 [4], 2013 AES Table-J [17].

Table 1. Primary energy consumption growth rate by fuel type in Australia.

Fuel type	Production (pJ) 2011-12	Growth (%)		Share (%)
		2010-11 to 2011-12	5 years average	
Coal	2118	−4.7	−2.3	34.2
Oil	2411	8.5	10.6	38.9
Gas	1399	4.2	1.2	22.6
RE	265	−7.3	-2.8	4.3
Total	6194	2.0	2.7	100

Sources: 2013 Australian Energy Statistic Data, Table C [17].

The coal use decreased because the falling coal use in iron and steel sector over the past five years. The renewable energy account for only 4.3% of total energy consumption. Among the renewable energy resources, wind energy contributed only 5.3% of growth whereas solar energy contributed significant growth by 20% from 2010-11 to 2011-12 [6].

2.3. Energy Production by Renewable Energy Sources

Australia is not only gifted with abundant, high quality and diverse non-renewable energy sources but also has large, widely distributed renewable energy sources like wind, solar, geothermal, hydroelectric, ocean energy and bioenergy resources. Australia's renewable energy resources are largely undeveloped.

Figure 3 shows the energy production from renewable energy sources from 2002-03 to 2011-12. The total renewable energy production reached a peak in 2007-08 and after that it gradually went down reaching a minimum in 2008-09. Now the trend is growing and rising. Australian's renewable energy consumption is 265 pJ which is share around 4.3% of the total energy consumption. The summary of renewable energy consumption, growth and share are presented in **Table 2**.

Figure 4 shows the Australia's total energy consumption by fuel type and their percent of share to the total consumption at a glance. The major contributions are from coal about 34.2%, and Oil 38.9%. So, coal and oil sectors contribution in Australian energy economy is more important.

2.4. Energy Consumption by Sectors

Now, it is needed to identify the sector which demands more energy consumption in Australia. The energy consumption by sectors like electricity generation, agriculture, mining, manufacturing and construction, transport,

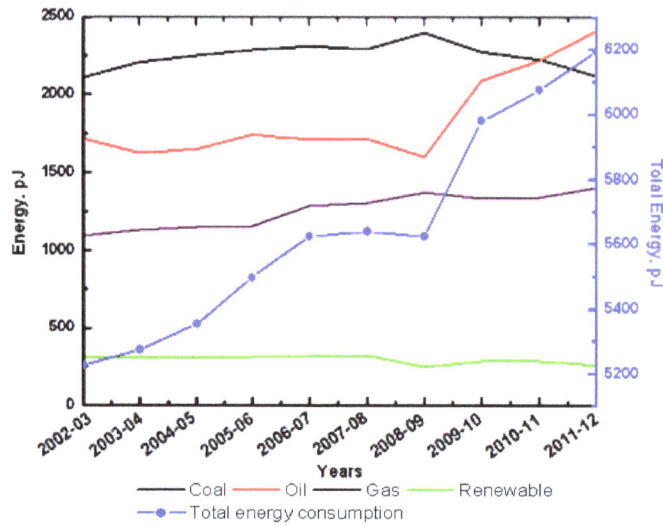

Figure 2. Australian's net energy consumption by fuel type [Unit 1 pe-
tajouls, pJ= 1015J]. Source: 2013 Australian Energy Statistics, Table C
[17].

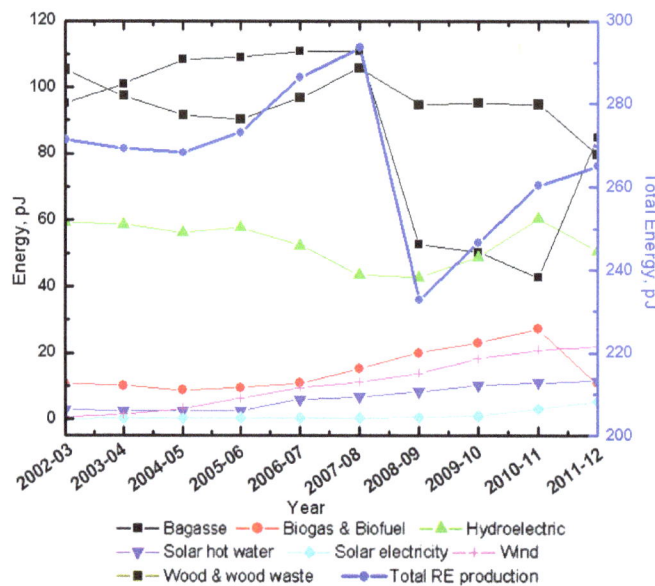

Figure 3. Energy production by renewable energy sources [Unit 1 pe-
tajouls, pJ= 1015J].Source: 2013 Australian Energy Statistics, Table A
[17].

commercial, residential and others is shown in **Figure 5**. It can be clearly seen from the Figure that the major
energy using sectors are electricity generation, transport and manufacturing together account for around 76% of
Australia's energy consumption.

The transport sector accounts for the largest share of Australia's end use consumption. According to Bureau
of Resources and Energy Economy (BREE) estimation, energy consumption of transport increased by an average of 2.4 percent per year during 2000-01 to 2011-12 [6]. This sector consumes mostly petroleum energy. The
next largest energy consuming sectors are mining, residential, commercial and services sectors.

3. Prospect of Renewable Energy in Australia

Australia has abundant and diverse renewable energy sources with significant potential for future development.

Table 2. Australian renewable energy consumption.

Fuel type	Consumption (pJ) 2011-12	Growth (%) 2010-11 to 2011-12	Share (%)
Biomass	165	- 0.9	2.3
Biogas/fuel	11	-55.7	0.4
Hydro	51	-16.2	1.0
Wind	22	5.3	0.3
Solar	17	19.9	0.2
Total	265	-7.3	4.3

Sources: 2013 Australian Energy Statistic Data, Table C [17].

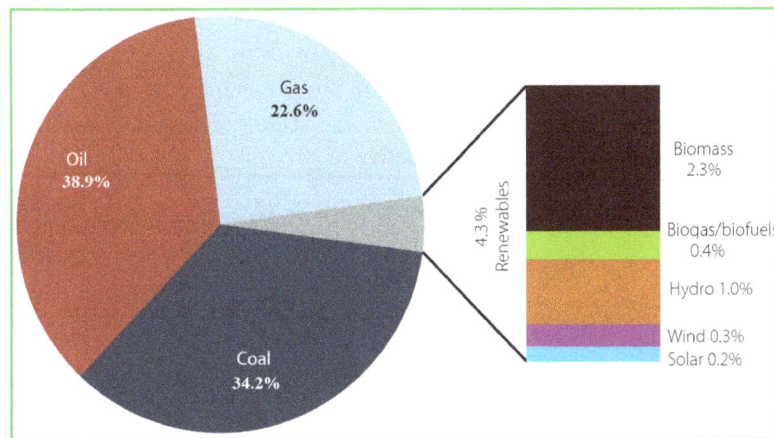

Figure 4. Total energy consumption and percent of shear. Source: 2013 Australian Energy Statistics, Table C [17].

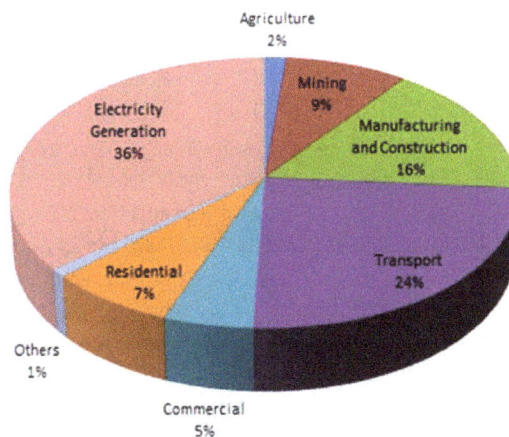

Figure 5. Total energy consumption by sectors (2011-12). Sources: 2013 Australian Energy Statistics, Table B [17].

Presently, renewable energy resources are used for heating and cooling, electricity generation, and as transportation fuels like bio-fuel. The clean energy resources being utilized on a commercial scale include hydro and wind energy for electricity generation, and bioenergy and solar energy for both heating and cooling and electricity generation [4]. Other renewable resources are mostly undeveloped at present and involve technologies which are still at the proof of concept stage or early stages of commercialization. A number of significant barriers are still being faced for large-scale utilization of Australia's clean energy resources. Changes in regulatory and approval processes are affecting well-established technologies like wind farms in some locations. It is often stated that the deployment of alternative energy resources will require a great deal of new research and development effort. The renewable energy technologies are too sophisticated and complex compared to conventional energy conversion technique. Due to the higher initial investment cost the renewable energy technology is not fully used yet. Despite these challenges, the deployment of clean energy technologies is gathering pace, and is expected to play a critical role in moving to a low emissions future while meeting Australia's continued demand for energy [4,6].

4. Conclusions

The following conclusions can be drawn from this review.

- Total energy production (which includes energy exports) in 2011-12 increased by 5%, relative to 2010-11, to total 17,460 petajoules, reflecting strong growth in natural gas (8%), black and brown coal (5% and 6%) and uranium (6%) production. Production of crude oil and liquefied petroleum gas fell 6%, and renewable energy decreased by 7% in 2011-12 compared to 2010-11.

- Total energy consumption, increased by 2%, relative to 2010-11, to total 6194 petajoules in 2011-12. The result was mainly driven by strong growth in energy use in the commercial and services sector and modest growth in the transport, mining, agricultural and residential sectors. Energy consumption in manufacturing and construction, however, decreased in 2011-12 relative to 2010-11.

- In 2011-12 consumption of renewable energy declined by 7%, relative to 2010-11, largely due to a fall in hydro energy consumption associated with lower hydroelectricity output in southeast Australia due to reduced water in-flows. Reduced hydro energy consumption more than offset the very strong growth observed in wind and solar energy.

- The prospect of renewable energy is more in Australia which will make a great contribution to Australia's energy economy in near future.

- The authors concluded that the Australia's renewable energy economic is viable; if its undeveloped renewable energy can be used effectively for electricity generation and transport sector.

References

[1]   Omri,A.(2013) CO$_2$Emissions, Energy Consumption and Economic Growth Nexus in MENA Countries: Evidence from Simultaneous Equations Models.*Energy Economics*,**40**, 657-664. http://dx.doi.org/10.1016/j.eneco.2013.09.003

[2]   Ong,H.C.,Mahlia,T.M.I. and Masjuki,H.H.(2011) A Review on Energy Scenario and Sustainable Energy in Malaysia.*Renewable and Sustainable Energy Reviews*,**15**, 639-647. http://dx.doi.org/10.1016/j.rser.2010.09.043

[3]   U.S.E.I. Administration (2013)International Energy Outlook. http://www.eia.gov/forecasts/ieo/

[4]   Willcock,C.N. andMcCluskey, C. (2013)Bureau of Resources and Energy Economics, Energy in Australia 2013. http://www.bree.gov.au/documents/publications/energy-in-aust/BREE-EnergyInAustralia-2013.pdf

[5]   Penney,A.S.K., Ball, A. and Hitchins,N. (2013)Bureau of Resources and Energy Economics, Energy in Australia 2012. http://www.bree.gov.au/documents/publications/energy-in-aust/energy-in-australia-2012.pdf

[6]   Che,A.F.N., McCluskey,C., Pham,P.,Willcock,T. and Stanwix,G. (2013)Bureau of Resources and Energy Economics. Australian Energy Update 2013. http://www.bree.gov.au/publications/aes.html

[7]   van Ettinger,J.(1994) Sustainable Use of Energy: A Normative Energy Scenario: 1990-2050.*Energy Policy*,**22**, 111-118. http://dx.doi.org/10.1016/0301-4215(94)90128-7

[8]   Masjuki, H.H.,Mahlia,T.M.I., Choudhury, I.A. and Saidur,R.(2002) Potential CO$_2$Reduction by Fuel Substitution to Generate Electricity in Malaysia.*Energy Conversion and Management*,**43**, 763-770. http://dx.doi.org/10.1016/S0196-8904(01)00074-7

[9]   Bushnell,J., Chen,Y. and Zaragoza-Watkins,M.(2014) Downstream Regulation of CO$_2$Emissions in California's Electricity Sector.*Energy Policy*,**64**, 313-323. http://dx.doi.org/10.1016/j.enpol.2013.08.065

[10]   Kirsten, S. (2014) Renewable Energy Sources Act and Trading of Emission Certificates: A National and a Supranational Tool Direct Energy Turnover to Renewable Electricity-Supply in Germany.*Energy Policy*,**64**, 302-312. http://dx.doi.org/10.1016/j.enpol.2013.08.030

[11]   Talaei,A.,Ahadi,M.S. and Maghsoudy, S.(2014) Climate Friendly Technology Transfer in the Energy Sector: A Case Study of Iran.*Energy Policy*, **64**, 349-363. http://dx.doi.org/10.1016/j.enpol.2013.09.050

[12]   Cohen,J.E.(2001) World Population in 2050: Assessing the Projections.*Conference on Series-Federal Reserve Bank of*

Boston, **2001**, 83-113.

[13] Vadiee,A. and Martin,V. (2014) Energy Management Strategies for Commercial Greenhouses.*Applied Energy*,**114**, 880-888.http://dx.doi.org/10.1016/j.apenergy.2013.08.089

[14] Hannam,P.(2013) Renewable Energy Study Tips Viable Reality by 2030. http://www.smh.com.au/business/carbon-economy/renewable-energy-study-tips-viable-reality-by-2030-201308232shb y.html#ixzz2l9LJldlH

[15] Puri,M.,Abraham, R.E.and Barrow, C.J. (2012) Biofuel Production: Prospects, Challenges and Feedstock in Australia.*Renewable and Sustainable Energy Reviews*,**16**, 6022-6031. http://dx.doi.org/10.1016/j.rser.2012.06.025

[16] D.O.E. Australian Government (2013)The Australian Energy Market Operator's 100 Percent Renewavle Study. http://www.climatechange.gov.au/reducing-carbon/aemo-report-100-renewable-electricity-scenarios/100-cent-renewab les-study-community-summary

[17] Bureau of Resources and Energy Economics (2013) Australian Energy Statistics Data. http://www.bree.gov.au/publications/aes-2013.html

18

Contribution of Vertical Farms to Increase the Overall Energy Efficiency of Urban Agglomerations

Podmirseg Daniel

Institute for Buildings and Energy Technical University, Graz, Austria
Email: daniel@podmirseg.com

Abstract

The 21st century keeps huge challenges for the system "city". Shortage of resources and world population growth forces architects to think in spaces with increasingly more structural linkages. No era has shaped the system of a city like the oil age did. Its grown structures are dependent from cheap and easy to produce petroleum. The postmodern city, facing the end of cheap and abundant oil, is now dependent from this finite resource. To minimize the dependency from hydrocarbon energy it is necessary to increase urban density, to switch to renewable energy production and to create new spaces for multifunctional purposes. An essential problem of urban agglomeration, though, is the fact that distances between food production and consumption have increased drastically in the last fifty years. Cheap oil made it possible to implement a global food transportation network and it also supported intensive monocultural food production. Today's food no more gets bought from local markets, but from labels. Its value is dependent from the brand-image, represented from the tertiary sector. The end of cheap fossil fuels carries a huge potential for architects and urban planners—we can move away from representing abstract, non-spatial processes and identities but creating spaces for dynamic local interactions. A promising typus for this might be the Vertical Farm.

Keywords

Vertical Farming; Energy Production; Energy Efficiency; Water Security

1. Biocapacity-Peak Oil

In the last fifty years world population growth entailed to a massive growth of urban agglomerations, not just in terms of density but also in land consumption. This development was also made possible through the production of massive amounts of petrochemical products. Low oil prices and easy production circumstances not just made these sturcural changes of cities possible, but by irony it made nowadays cities dependent on this situation. This fact is a huge challenge for the city of tomorrow.

"Peak oil" [1] and the upcoming explosion of prices of fossil fuels make it necessary to reconstruct and reconfigure the system "city".

Massive land use and easy-to-get oil also lead to actual systems of food production [2], especially in industrialized and emerging countries: Food no longer gets produced where it gets consumed [3]. A global food network emerged nowadays in mega cities is completely dependent and couldn't supply themselves with daily needed edible biomass.

The question whether the biocapacity of the earth is big enough to feed world population of the next generations actually is intensively discussed between scientists and researchers [4] [5]. Complex calculation models show different results and methodologically strongly different approaches make results even more difficult to compare.

But what we know is that a third of our earth's surface is covered with land mass. And a third of it has the capacity to produce food [6]. This area, globally, is unevenly allocated and the dependency of global food transportation networks is inherent. This view on the problem of food security of urban agglomerations demonstrates the challenge to increase the overall energy efficiency of cities, especially in food production.

2. What Is a Vertical Farm?

The idea of food production in city indeed is not new. Glass- and greenhouses, courtyard-gardening, rooftop gardens and winter gardens are common elements of cities. On an inflationary way low- and high-tech experiments get published for years now; with very little potential of relevant energy production (directly or via biomass). But a closer look to these enterprises envision, beside the economic advantages (which in most cases can get ignored), a unique side-effect: a new spatialization of direct interdependencies between different interests and professions [7] [8].

The Vertical Farm, although, on an industrial level, promises the biggest efficiency in terms of biomass-production/m^2 [9] [10].

Cultivation area gets STACKED, CONDITIONED and URBANIZED: These three headwords outline the principle of Vertical urban Farms.

3. Stacking

Actually every inhabitant, if he's an omnivore, needs ca. 2300 m^2 [11] on cultivation area to produce enough fruit, vegetable, wheat, oil seeds, semi luxury food, milk, meat and fish, but also feed for animal husbandry.

Research results show that food production (especially vegetable and fruit) in a controlled environment can radically reduce the needed cultivation area (up to thirty times) and the ecological footprint in food production [12].

This allows an intensification of biological or alternative food production like permaculture on areas Vertical Farms are compensating [13]. Giving agricultural areas, damaged by conventional cultivation practises, back to nature, soil can be regenerated or reforestation is thinkable.

4. Conditioning

In a conditioned stacked greenhouse edible biomass production can be guaranteed 365/24hours a day [14]. This excludes food losses and crop reductions through droughts, hail and other weather conditions. Technologies like hydroponics and aeroponics obviate the use of herbicides, fungicides and fertilizers [15]. Another advantage offers the closed water cycle which guarantees a drastic reduction of water use in food production and enables the possibility of water recycling and the reuse of grey water [16].

Simulation- and calculation models although also show that the provision of the necessary light intensity to support photosynthesis leads to relevant energy consumption [17].

5. Urbanization

Actual system borders of the basic needs-field "nutrition" only can be analysed on a global scale [18]. Material flow analysis [19] of this basic needs-field is more and more the primary subject of different scientific research works, mostly compared with a CO_2-balance of different food products. These results envision an enormous potential to increase the overall energy efficiency of urban agglomerations by implementing Vertical Farms.

Numerous mega cities import their food up to 80% not from the surrounding of the city border or from the country itself, but from foreign countries [20]. Food miles as a consequence are related to massive energy consumption (through fossil fuels) and lead to drastic CO_2-output [21].

Furthermore food miles lead to direct and indirect costs on a social and economic level, and on the environment. This situation is one of the most important reasons why the ecological food footprint [22] multiplies the per capita food production footprint per five times [23]—up to 15,000 m².

Especially for architects and urban planners these circumstances offer an opportunity to examine and develop the typus of the Vertical Farm. The functional system borders of conventional agriculture can be described as follows [24]:

Seeds, fertilizers, pesticides and herbicides have to be produced before they reach the cultivation area via transport. After an energy intense cultivation and animal husbandry-period products, again through transport, get provided or to food processing industries or wholesale trade systems (storage). In this moment huge amounts of energy are needed for cooling systems, storage, packaging materials and packaging processes, before, again via transport, food products get distributed to markets, households or again to decentralized storages [25].

These processes have to be closed in a system in Vertical Farms. Production of nutrition substances, animal husbandry and edible biomass cultivation, food processing and waste management (eventually for energy production), water management and implementation of installation technology (hvac) and lighting find place on a local scale. This reimplementation of local social and economic interdependencies in this context can be read as the biggest potential of Vertical Farms.

Material flow analysis—the comparison from conventional agriculture and the Vertical Farm is the nucleus of the dissertation. The essential question comes up is: To what extend can the Vertical Farm be a contribution to increase the overall energy efficiency [26] of urban agglomerations? What is the difference between the saved energy through shrinking of system borders of food production [27] on a local level and the energy needed to run a Vertical Farm?

Especially for the Symposium questions in this context should change their focus on spatial potential and qualities to enable discussions on architectonic and urbanistic levels [28].

What are the potential spatial intersections between the production entity and the public space? Which new concepts of public and semi-public spaces can be envisioned by implementing hybrid types (Vertical Farm-Housing, Vertical Farm-Shopping Malls, Hyperbuildings···)? Which additional functions only can lead to a hybridization (market, trade, multifunctional horizontal/vertical public spaces, parks, spaces for social gathering)?

References

[1]   (2012) Oil&Gas2_Campbell update 2012.xlsx

[2]   (2008) The Atlas of Food, Millstone.

[3]   (2008) The Atlas of Food, Millstone.

[4]   (2000) Global Agro-Ecological Zones Assessment, IIASA.

[5]   (2010) Towards the Third Green Revolution, Harald von Witzke.

[6]   (2009) World Resource Outlook to 2050, Jette Bruinsma.

[7]   (2006) The Power of Community, How Cuba Survived Peak Oil.

[8]   (1996) Local Food Systems and Sustainable Communities, Feenstra.

[9]   (2010) The Vertical Farm, Dickson Despommier.

[10]   (2004) Advanced Life Support Baseline Values and Assumptions Document, NASA.

[11]   (2012) Tonnen für die Tonne, WWF und Harald von Witzke.

[12]   (2012) Market Analysis for Terrestrial Application of Advanced Bio-Regenerative Modules: Prospects for Vertical Farms, Chirantan Banerjee.

[13]   (2009) Farm of the future, produced by Tim Green and Rebecca Hosking, BBC.

[14]   (2010) The Vertical Farm, Dickson Despommier.

[15]   (2010) The Vertical Farm, Dickson Despommier.

[16]   (2004) NASA Advanced Life Support Baseline Values.

[17]   (2012) Market Analysis for Terrestrial Application of Advanced Bio-Regenerative Modules: Prospects for Vertical Farms, Chirantan Banerjee.

[18]   (2005) Umweltauswirkungen von Ernährung-Stoffstromanalysen und Szenarien, Wiegmann/Eberle/Fritsche/Hünecke.

[19] (2008) CO$_2$-Bilanz der Tomatenproduktion: Analyse acht verschiedener Produktionssysteme in Österreich, Spanien und Italien, Theurl.

[20] (2002) City Limits, Livingston.

[21] (2005) DEFRA—Department of Environment Food and Rural Affairs, Validity of Food Miles UK.

[22] (2005) DEFRA—Department of Environment Food and Rural Affairs, Validity of Food Miles UK.

[23] (2009) Time to Eat the Dog? Robert and Brenda Vale.

[24] (2005) Umweltauswirkungen von Ernährung-Stoffstromanalysen und Szenarien, Wiegmann/Eberle/Frische/Hünecke.

[25] (2005) Umweltauswirkungen von Ernährung-Stoffstromanalysen und Szenarien, Wiegmann/Eberle/Frische/Hünecke.

[26] (2010) The Vertical Farm, Dickson Despommier.

[27] (2000) Ressourceneffizienz in der Aktivität Ernährung, Mireille Chloé, Jeanne Rachel Faist, ETH Zürich.

[28] (2012) DIO NERO, from Citigroup to Citicrop Headquarter, Exhibition, IMDP.

19

Shape Effect of Piezoelectric Energy Harvester on Vibration Power Generation

Amat A. Basari[1], Sosuke Awaji[1], Song Wang[1], Seiji Hashimoto[1], Shunji Kumagai[2], Kenji Suto[2], Hiroaki Okada[2], Hideki Okuno[2], Bunji Homma[2], Wei Jiang[3], Shuren Wang[3]

[1]Division of Electronics and Informatics, Gunma University, Kiryu, Gunma, Japan
[2]Research and Development Department, Mitsuba Corporation, Kiryu, Gunma, Japan
[3]School of Hydraulic, Energy and Power Engineering, Yangzhou University, Yangzhou, China
Email: hashimotos@gunma-u.ac.jp, s-kumaga@mitsuba.co.jp, jiangwei@yzu.edu.cn

Abstract

Vibration energy harvesting is widely recognized as the useful technology for saving energy. The piezoelectric energy harvesting device is one of energy harvester and is used to operate certain types of MEMS devices. Various factors influence the energy regeneration efficiency of the lead zirconate titanate piezoelectric (PZT) devices in converting the mechanical vibration energy to the electrical energy. This paper presents the analytical and experimental evaluation of energy regeneration efficiency of PZT devices through impedance matching method and drop-weight experiments to different shape of PZT devices. The results show that the impedance matching method has increased the energy regeneration efficiency while triangular shape of PZT device produce a stable efficiency in the energy regeneration. Besides that, it becomes clear that the power, energy and subsequently efficiency of the triangular plate are higher than those of the rectangular plate under the condition of the matching impedance and the same PZT area.

Keywords

Vibration Power Generation, PZT Device, Impedance Matching, Energy Regeneration Efficiency

1. Introduction

In recent years, solutions for the environmental and energy problems correspond to the increment of power demand all over the world have become a great interest of research. One of the solution methods that highly recommended by literature is by doubling our efforts in research and development on renewable energy technology. Wind, solar, thermal, vibration and many other types of energies are always available and they do not adversely affect the environment. To date, for example, solar, thermal and wind energy have successfully been converted to usable electrical energy and used in various industrial and home appliance products [1]-[3], while vibration energy has yet to become an alternative source of energy for self-powered system. Conversion of vibration to electrical energy is possible using piezoelectric, electromagnetic and electrostatic based devices [4]. It is stated in [5] that, generally, electrostatic device is able to produce electrical energy up to 2% of efficiency while electromagnetic device according to [6], it can generate electrical power up to 1.4 mW with 25% efficiency. A lot of efforts are being introduced to optimize the energy output of every device mentioned in above. In case of piezo-electric device, analytical analysis on the vibration energy regeneration efficiency of the devices is proposed by Adachi *et al.* in [7]. Kong *et al.* in [8] presents a resistive impedance matching circuit for vibration energy harvesting. It is proven that impedance matching circuit can be utilized for optimization of energy conversion. In [9] the authors work on sensor shape design and in [10] structure of the device has been focused for the optimum energy output. Another factor that can contribute to the variation in the output level of the vibration energy harvesting system is the power conditioning circuits. This topic has been discussed in [11] and a detail summary can be found in [12].

With the recent surge of micro scale devices, piezoelectric power generation can provide a convenient alternative to traditional power sources used to operate certain types of sensors/actuators and MEMS devices. However, the energy produced by these materials is in many cases far too small to power an electrical device.

This paper presents the evaluation of vibration energy regeneration efficiency of PZT devices which is produced through impedance matching approach and by different shape of PZT devices. In addition to these, energy regeneration efficiency of PZT devices under the same PZT area and matching impedance is also evaluated analytically and experimentally.

2. Impedance Matching of PZT Device

2.1. Power and Energy Generation with PZT Devices

Impedance matching is one of the methods to increase power and energy output of PZT device power generator. Theoretically, by matching the load impedance with input impedance, maximum power can be delivered from source to the load. Prior to the evaluation on the impedance matching of the four selected bimorph PZT devices, discussion on power and energy generation by all PZT devices will be presented.

To evaluate the power and energy generation by all PZT devices, experimental setup as shown in **Figure 1** was built. The experiments were conducted based on the experimental conditions as listed in **Table 1**. Relation

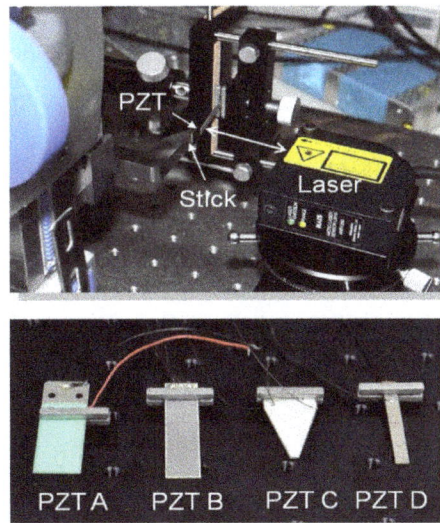

Figure 1. Experimental setup and PZT devices.

Table 1. Experimental conditions.

Parameters	Details
1) Input signal	Step signal
2) Amplitude	0.5 mm
3) Measurement signals	PZT voltage and displacement of PZT device's free end
4) Sampling time	0.2 ms
5) Displ. resolution	3.5 μm
6) ADC resolution	16 bit

ship between the resonant frequency, matching impedance and energy is described in the following section.

2.2. Impedance Matching for Maximum Power and Energy Generation

In the previous section, discussion on power generation by the PZT devices with fixed load resistor of 10 kΩ has been presented. This section will discuss the impedance matching of PZT devices so that maximum power and energy can be generated. For PZT device A, load resistor was increased gradually from 100 Ω to 3 MΩ. The results of power dissipation by each resistor are shown in **Figure 2**.

From the figure, maximum power was dissipated by the 12.8 kΩ resistor and maximum energy was generated when resistor is 100 kΩ. Notice that the maximum power and energy are dependent on different value of resistors. From this observation we know that PZT device A has two resonant frequencies.

Next, power spectral density (PSD) of 100 kΩ resistor was plotted. It is clearly can be seen in **Figure 3**, the resonant frequency of PZT device A appears at two different frequencies; at 39 Hz and 586 Hz. For PZT device B, C and D, same experiments were conducted and their dissipated voltage's power spectral densities were analyzed. Maximum power and energy for each PZT device were found to be dependent on the load resistors.

2.3. Energy Regeneration Efficiency

To evaluate the energy regeneration efficiency of each PZT device, experimental set up as shown in **Figure 4** was constructed and ratio of output to input energy was calculated. Note that one end of the cantilever beam was clamped and the other end was free and attached with mass $m = 0.1$ kg by string. At the rest position, the displacement of the free end is x_0. The displacement of the free end after the string is cut was measured using laser measuring device while the voltage drop was measured using voltmeter.

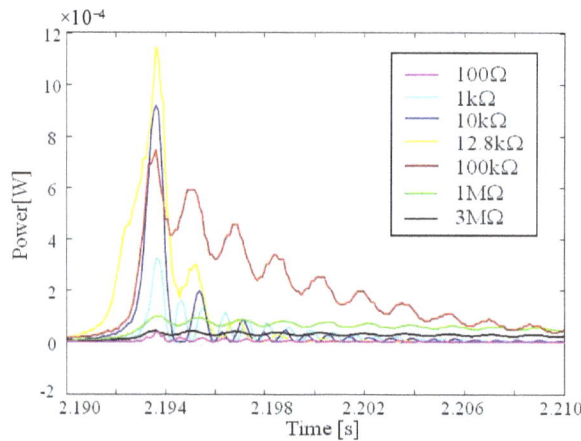

Figure 2. Power dissipation by resistors—PZT device A.

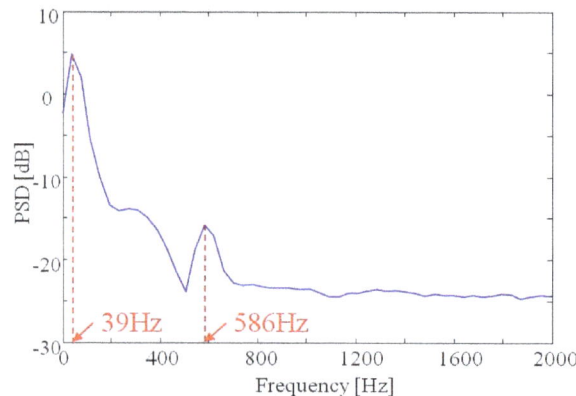

Figure 3. Voltage power spectral density of 100 kΩ resistor.

Without considering the impedance matching results, the load resistor is set to 10 kΩ for all PZT devices. Input energy, output energy and energy efficiency were calculated based on the Equations (1)-(3). Results are summarized in **Table 2**.

$$W_i = \int F dx \approx \frac{1}{2} mgx_0 \tag{1}$$

$$W_o = \frac{1}{R} \int V^2 dt \tag{2}$$

$$\eta = \frac{W_o}{W_i} \times 100\% \tag{3}$$

Now, based on the impedance matching results, load resistor was replaced with the impedance matching resistor that produced maximum energy for each PZT device. The displacement and voltage drop were measured and recorded. As shown in **Table 2**, it is clear that energy regeneration efficiency has increased between 4% to 88% for all PZT devices with the impedance matching resistor.

3. Energy Regeneration Efficiency of Different Shape PZT Devices

This section will discuss the effect of the shape of PZT device to the energy regeneration efficiency. In the previous section, four different size PZT devices have been tested. Basically, PZT device A, B and D have the same shape. Thus, in this section, to evaluate the effect of the PZT shape on the energy regeneration efficiency, PZT device A and C which having different shapes were used.

3.1. Stress, Deflection and Elastic Energy Analysis

Dimension of PZT device C in the cantilever beam structure is shown in **Figure 5**. In this figure, W represents the concentrated load that applied on the free end of the cantilever beam. Theoretically, in bending mode, maximum stress is calculated using Equation (4).

$$\sigma_{max} = \frac{M}{Z} = -\frac{6W_x}{bh^2} \tag{4}$$

where, M is bending moment, Z is section modulus, x is distance from the load, b is the width and h is the thickness of the PZT plate. From Equation (4), thickness h is a fixed value, and if ratio of x/b is also fixed, the maximum stress at all points of the PZT plate will become constant. Based on this fact, different from PZT device A which having the maximum stress at the clamped point, PZT device C is the one with the constant value of maximum stress at all points of the plate for which the width b will increase as distance from the free end x in

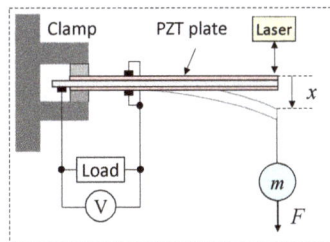

Figure 4. Experimental setup for energy measurement.

Table 2. Energy regeneration efficiency.

PZT	Efficiency with 10kΩ resistor [%]	Efficiency with impedance matching resistor [%]	Increment in efficiency
A	3.0	5.4	80%
B	4.8	5.0	4%
C	3.1	4.9	58%
D	1.7	3.2	88%

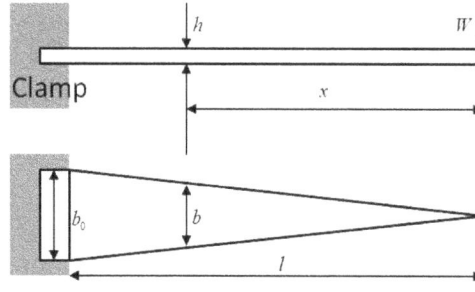

Figure 5. Cantilever beam of PZT device C.

creases.

Further analysis on deflection and elastic energy experienced by both PZT devices was performed. Based on the above analysis, the curvature of PZT device C will be a constant value when concentrated load is applied. Moreover, if compared to PZT device A, the deflection angle of the PZT C is double and this will increase its deflection and the input energy W_i, to 1.5 times of that the PZT device A has. The equation that denoted the input energy is shown in equation 1. By looking at the equation, it can be expected that energy regeneration efficiency will become better than what PZT device A has. Another merit point that PZT device C has here is if we look to its volume. By having the same durability, the cost also can be reduced as its volume is half of the PZT device A.

Meanwhile, in terms of elastic energy per unit volume for PZT device A and C, they are denoted by Equations (5) and (6) respectively.

$$U_A = \frac{1}{2}Wy_0\frac{1}{V_A} = \frac{\sigma^2}{18E} \tag{5}$$

$$U_C = \frac{1}{2}Wy_0\frac{1}{V_C} = \frac{\sigma^2}{6E} \tag{6}$$

where y_0 is the maximum deflection, E is the Young's modulus and V is the volume. As mentioned in above, with 1.5 times of maximum deflection and half of the volume, it is clearly can be seen from the equation that PZT device C will be able to absorb 3 times greater externally applied elastic energy than of that the PZT device A would absorb.

3.2. Experimental Analysis of Shape Effect

Energy regeneration by PZT device is very much related to the stress that can be generated by the PZT device where we know that deflection is the source for the stress. An increment in the deflection simply can be done by increasing the concentrated load.

Therefore, to observe the effects of the shape of PZT plate to the energy regeneration efficiency, an experimental setup as shown in **Figure 4** was constructed. The experiments were conducted with load mass which was used as the varying parameter. It was increased gradually from 50 g to 350 g. The output was connected to the impedance matching resistor of each case. Input and output energy data were measured and recorded.

From **Figure 6**, it can be seen that both input energy of PZT device A and C increase with quadratic function of the load mass. Besides that, the output energy of PZT device C, as shown in **Figure 7** also increases with the same function. Contradict to the rest, the output energy generated by the PZT device A is saturated at low level even the load mass is increased. In terms of energy regeneration efficiency, as can be seen in **Figure 8**, the PZT device C marked a stable efficiency at about 5% as the load mass increases when compared to that of the PZT device A.

3.3. Shape Effect on Same PZT Area

To experimentally evaluate the shape effects, as illustrated in **Figure 9**, PZT devices were attached on the same size of rectangular and triangular plates. The length a, b and c are 59, 30 and 10 mm, respectively. Experimental analyses with step signal input were first carried out and matching impedances for maximum power and energy

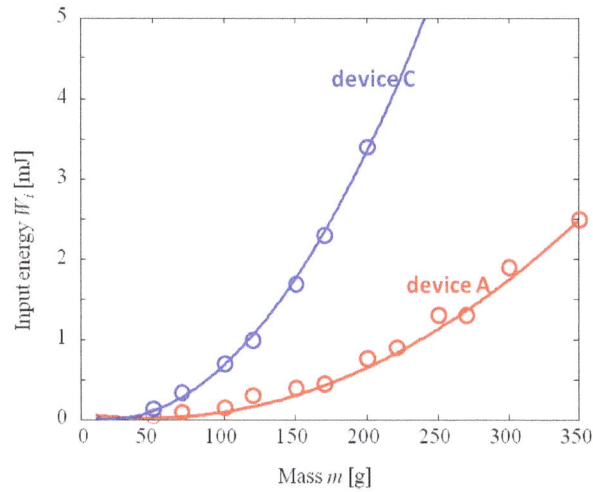

Figure 6. Input energy of PZT cantilever beam versus load mass.

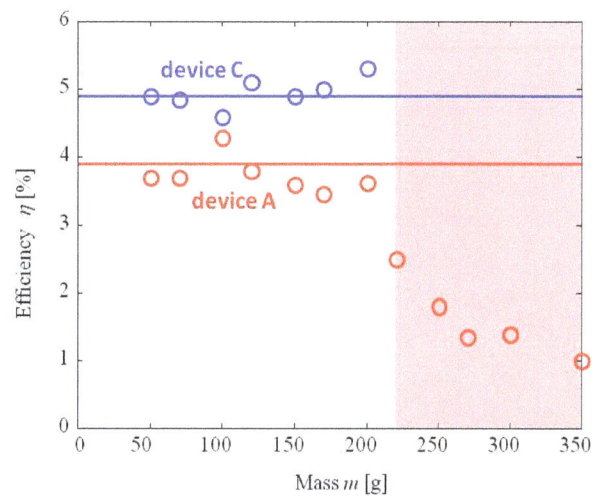

Figure 7. Output energy of PZT cantilever beam versus load mass.

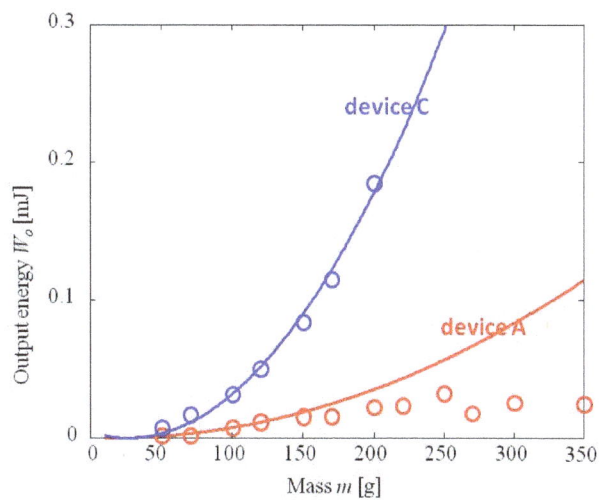

Figure 8. Energy regeneration efficiency of PZT cantilever beam versus load mass.

generation were derived. Initially, free end of the plates were deflected by 3 mm from its origin and then, displacement of the free end and voltage across the load were measured.

The results are as follows; For the triangular plate, resonant frequencies of 185 Hz and 372 Hz generate maximum energy and power with 4 kΩ and 2 kΩ matching impedances respectively. On the other hand, the resonant frequency for the rectangular plate is 104 Hz and its matching impedance is 10 kΩ. Since the resonant frequencies of the triangular plate are higher than that of the rectangular plate, the matching impedances for the triangular plate are smaller than that for the rectangular plate. For the triangular plate, two resonant frequencies were measured. As a result, two matching impedances are obtained.

Next, by using the obtained matching impedances, energy generation efficiency of the plates were experimentally investigated. This time, mass was attached by string to the free end of the plates and after the string was cut, its displacement and instantaneous voltage across the load were measured. The input and output energy for each mass can be calculated from (1) and (2). **Figure 10** and **Figure 11** show the input and output energy plots respectively. It can be seen from the plots that both curves are increase quadratically with mass.

Moreover, the calculated efficiency by (3) is shown in **Figure 12**. From **Figure 12**, since the resonant fre-

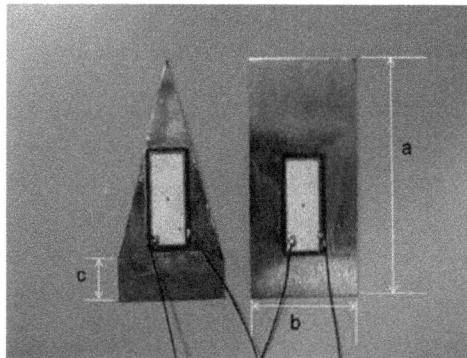

Figure 9. Picture of triangular and rectangular plates.

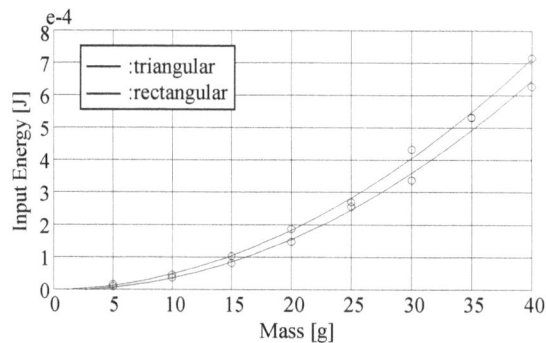

Figure 10. Input energy versus load mass.

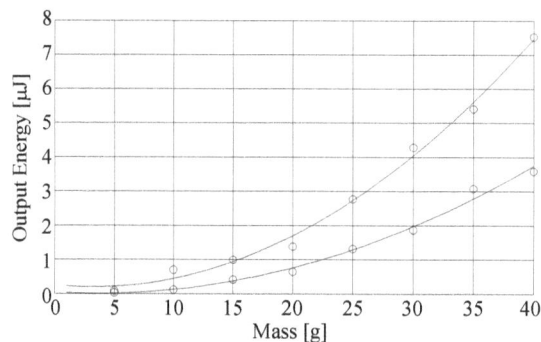

Figure 11. Output energy versus load mass.

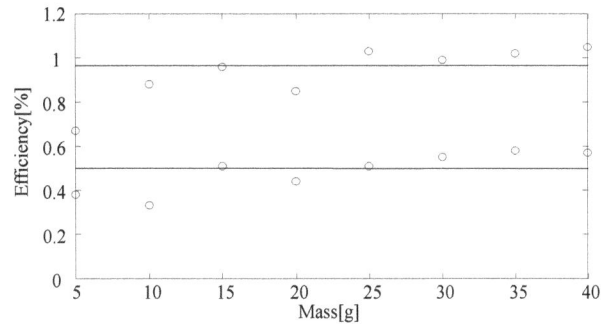

Figure 12. Efficiency comparison.

quency of the triangular plate is higher than that of the rectangular plate, the matching impedances are smaller. As a result, current, power, energy and subsequently efficiency of the triangular plate are higher than that of the rectangular plate.

4. Conclusion

This paper has successfully evaluated how impedance matching, shape of devices, concentrated mass and forced vibration contribute to the variation of the energy regeneration efficiency of PZT devices. Input and load impedance matching has shown an increment in the energy regeneration efficiency to between 4% and 88%. On the other hand, a stable 5% energy regeneration efficiency was produced by a triangular plate of PZT device while the rectangular plate's efficiency has worsened as load mass increased.

Further analysis on the output of triangular PZT device shows that power, energy and subsequently efficiency of the triangular plate are higher than those of the rectangular plate under the condition of the matching impedance and the same PZT area.

References

[1]   Rizza, J.J. (2013) Solar-Driven LiBr/H_2O Air Conditioning System with a R-123 Heat Pump Assist. *Journal of Solar Energy Engineering*, **136**, 1-5. http://dx.doi.org/10.1115/1.4024741

[2]   Beshore, D.G., Jaeger, F.A. and Gartner, E.M. (1979) Thermal Energy Storage/Waste Heat Recovery Application in the Cement Industry. *Proceedings of First Industrial Energy Technology Conference*, 747-756.

[3]   Hatziargyriou, N. and Zervos, A. (2004) Wind Power Development in Europe. Proc. of the IEEE, 1765-1782.

[4]   Hulst, R.D., Sterken, T., Puers, R., Deconinck, G. and Driesen, J. (2010) Power Processing Circuits for Piezoelectric Vibration-Based Energy Harvesters. *IEEE Transaction on Industrial Electronics*, **57**, 4170-4177. http://dx.doi.org/10.1109/TIE.2010.2044126

[5]   Miranda, J.O.M. (2004) Electrostatic Vibration-to-Electric Energy Conversion. PHD thesis, Massachusetts Institute of Technology.

[6]   Robert, G. and Radu, O. (2011) Harvesting Vibration Energy by Electromagnetic Induction. Electrical Engineering Series, 7-12.

[7]   Adachi, K. and Sakamoto, T. (2012) Study on Energy Transfer Efficiency Analysis of Cantilever Type of Piezocomposit Vibration Energy Harvester. The Japan Society of Mechanical Engineers, 271-281.

[8]   Friswell, M.I. and Adhikari, S. (2010) Sensor Shape Design for Piezoelectric Cantilever Beams to Harvest Vibration Energy. *Journal of Applied Physics*, **108**, 1-7.

[9]   Hashimoto, S., Nagai, N., Fujikura, Y., Takahashi, J., Kumagai, S., Kasai, M., Suto, K. and Okada, H. (2012) Multi-Mode Vibration-Based Power Generation for Automobiles. *Proceedings of the* 2012 *IEEE-IAS Annual Meeting*, 1-5.

[10] Young, M. (1989) The PWM Strategy on DC-DC Converter. *IEEE Journal of Industry Applications*, **28**, 123-129.

[11] Eason, G., Noble, B. and Sneddon, I.N. (1995) On Certain Integrals of Lipschitz-Hankel Type Involving Products of Bessel Function. *IEEE Transaction on Power Electronics*, **247**, 529-551.

[12] Maxwell, J.C. (2010) A Treatise on Electricity and Magnetism. *IEEE Transaction on Industry Applications*, **589**, 68-73.

20

Prospects of Renewable Energy in Semi-Arid Region

K. S. V. Swarna, G. M. Shafiullah, Amanullah M. T. Oo, Alex Stojcevski

School of Engineering, Faculty of Science, Engineering and Built Environment, Deakin University, Geelong Waurn Ponds Campus, Australia
Email: ssrungar@deakin.edu.au

Abstract

Continuous usage of fossil fuels and other conventional resources to meet the growing demand has resulted in increased energy crisis and greenhouse gas emissions. Hence, it is essential to use renewable energy sources for more reliable, effective, sustainable and pollution free transmission and distribution networks. Therefore, to facilitate large-scale integration of renewable energy in particular wind and solar photovoltaic (PV) energy, this paper presents the feasibility analysis for semi-arid climate and finds the most suitable places in North East region of Victoria for renewable energy generation. For economic and environmental analysis, Hybrid Optimization Model for Electric Renewables (HOMER) was used to investigate the prospects of wind and solar energy considering the Net Present Cost (NPC), Cost of Energy (COE) and Renewable fraction (RF). Six locations are selected from North East region of Victoria and simulations are performed. From the feasibility analysis, it can be concluded that Mount Hotham is one of the most suitable locations for wind energy generation while Wangaratta is the most suitable location for solar energy generation. Mount Hotham is also the best suitable locations in North East region for hybrid power systems i.e., combination of both wind and solar energy generation.

Keywords

Renewable Energy; HOMER; Semi-Arid; Feasibility; Optimization; Sensitivity

1. Introduction

Most of the electrical energy generated in Victoria is by burning coal and from other fossil fuels as it has substantial reserves of conventional resources. As a result, environmental issues have become major concern due to the continuous emission of carbon dioxide in to the atmosphere causing global warming. Renewable generation is more efficient as it is obtained from sources that are reliable and inexhaustible and does not release any greenhouse gas (GHG) or toxic gases when producing electricity [1-4]. Hence, renewable energy technology is an optimum solution that has a much lower environmental impact than other conventional energy technologies [5,6]. In 2007, Australian government ensured that by 2020 the share of renewable energy in total electricity generation should be of 20%. Hence, integration of Renewable energy sources is required forming a hybrid energy system with better efficiency and cost of energy. Victoria's total renewable generation for 2012 was around 3825 Giga-watt hours which is nearly 30% more than the generation in 2011 [7,8]. Wind generation is one of the fastest growing pollution free technology and cost effective among different renewable energy generations [9,10]. Solar PV has also experienced rapid growth driven largely by house hold systems [5].

This paper investigates the prospects of renewable energy (RE) sources and identifies the most suitable places of semi-arid region in particular North East region in Victoria for wind and solar considering economical performance metrics COE and NPC [11,12]. This paper also focuses on emission analysis by considering the performance metrics RF and GHG emission [13]. Details of these performances metrics are available in [14,15]. This study developed hybrid models with HOMER and conducted sensitivity analysis with different probabilities to know the suitable options considering different meteorological conditions. This paper is organized in the following manner: Section II describes the Hybrid Renewable Energy Model, Section III gives the complete optimization and sensitivity analysis and finally section IV concludes the paper with future analysis.

2. Hybrid Renewable Energy System

Hybrid power systems combine two or more energy conversion mechanisms, or two or more fuels for the same mechanism, that when integrated, overcome limitations inherent in either. Hybrid systems provide a high level of energy security and reliability through the integrated mix of complementary generation methods, and often will incorporate a storage system (battery, fuel cell) or fossil-fueled power generation to ensure consistent supply [16,17]. This section concentrates on system modeling, energy resources used, electric load, standard grid and the converter. In this analysis a hybrid energy system is designed with an integration of solar and wind power generation system. **Figure 1** shows the Hybrid energy system modeling in HOMER with solar and wind resources. In this model, a separate storage device is not used for ease of operation as the grid itself acts as a storage device. The following sections gives the additional information for load, renewable energy resources and the other components used in the hybrid power systems.

2.1. Elecric Load

The average load profile considered in this analysis is on the Victorian monthly load demand. Scaled annual average load for the current system is 1621 kWh/day and the annual peak load is 178 kWh. **Figure 2** shows the average load profile and it is observed that the highest load demand is seen in the month of November.

2.2. Renewable Energy Resources

Data have been collected from Bureau of Meteorology (BOM) [18] for North East region and based on the maximum annual solar irradiance three best locations are chosen for PV generation and based on maximum annual wind speeds other three locations are selected for wind generation. Therefore, in total six locations are identified from North East region and simulations are performed with HOMER. **Table 1** shows the annual average wind speed and solar irradiance of six locations considered in the study.

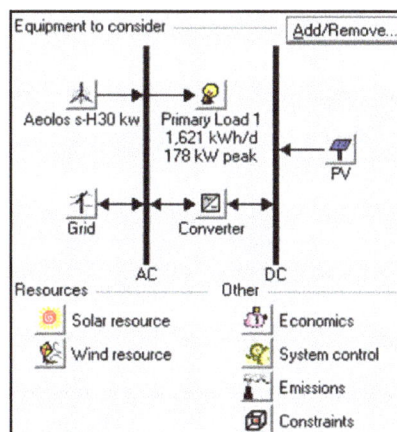

Figure 1. Hybrid renewable energy system with HOMER.

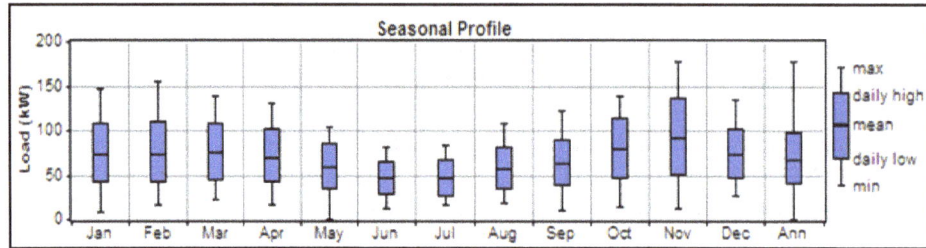

Figure 2. Average load profile for a year.

Table 1. Annual average of wind and solar in six regions.

Location	Annual average	
	Wind (m/sec)	Solar (kWh/m^2/)
Mount Hotham	6.9	4.3
Falls creek	6.3	4.3
Mount Buller	5.6	4.3
Wangaratta	1.7	4.8
Edi upper	2.7	4.7
Benalla	2.2	4.3

2.2.1. Wind Energy Resource

From the **Table 1** it is clear that the wind speed is maximum for the Mount Hotham area which is around 6.9 m/sec followed by Falls Creek and Mount Buller. **Figure 3** demonstrates the monthly wind speed variations of Mount Hotham with maximum annual average and it is observed that the wind energy is maximum in summer from the month of September to March.

2.2.2. Solar Energy Resource

HOMER uses the daily solar resource inputs to calculate the PV array power for each hour of the year. From the **Table 1,** it can be observed that Wangaratta location has maximum annual solar irradiance of 4.8 kWh/m^2/day. **Figure 4** shows the clearness index and its value is varied from 0.46 to 0.64 and the maximum solar radiation observed is in the month of December.

2.3. Component Details

The essential components for this model are Photovoltaic modules, wind turbine, converter and standard grid.

2.3.1. Photovoltaic

The capital cost for a 1.0 kW photovoltaic array considered in this analysis is $2100 with a replacement cost of $2000. For optimum solution, the O&M cost is practically considered to be zero. Sizes of photo voltaic cells to be considered are from 0 to 250

2.3.2. Wind Turbine

For simulation Aeolos s-H30, 30 kW wind turbine is considered with a capital cost of $66,000 [19] and with zero O&M. The sizes to consider during this analysis are 10, 12 and 14. **Figure 5** shows the power curve for the assumed wind turbine.

2.3.3. Power Converter

Converter serves the purpose of both rectifier and inverter. The DC energy from the photovoltaic is converted into AC using an inverter. For 1 kW of energy, the installation cost is considered to be $800 and the replacement cost is assumed to be $750. The life time is assumed to be 20 years considering an efficiency of 90%.

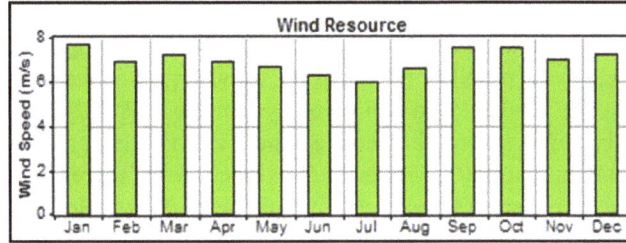

Figure 3. Wind variations for Mount Hotham region.

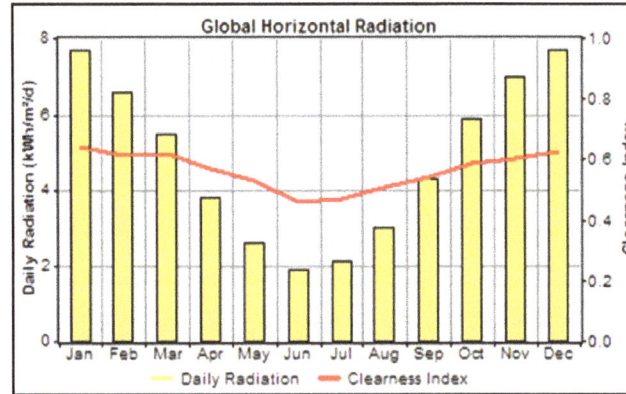

Figure 4. Average solar irradiance for Wangaratta.

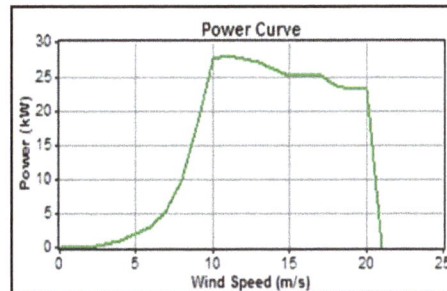

Figure 5. Power curve for Aeolos s-H30 wind turbine.

2.3.4. Grid

Grid acts as backup component supplying energy to meet the demand load during surplus energies. Hence in this model separate battery is not used for storage and the grid itself is acting as a storage device [20]. A standard grid price of 0.4 $/kWh is maintained throughout the study.

3. Result Analysis

This study provides the results in terms of optimization and sensitivity analysis using HOMER simulation tools. Optimization results are important as this provides the economic feasibility and environmental friendly information under specific meteorological conditions. Whereas, sensitivity analysis is a measure that checks the sensitivity of a model while changing the value of the parameters of the model which helps in decision making. This model has been simulated and analyzed based on the sensitivity variables wind speed, solar irradiation, and grid electricity price.

3.1. Optimization Results

The data collected from BOM is given as inputs to HOMER model and is used for calculating the NPC, COE and RF. Simulation results are calculated for the six locations as stated in **Table 1**. From **Figure 6**, it is observed that NPC and COE for Mount Hotham are reduced to 55% with the integration of wind energy into the grid in place of only grid connected systems.NPC, COE and RF of wind/grid connected system are $1,649,920, 0.218 $/kWh and 87% while in the wind/PV/grid connected system are $1,844,260, 0.224 $/kWh and 85% respectively. From **Figure 7** it has been evident that from a wind/PV/Grid connected systems average electricity production from wind, PV and grid are 80%, 5% and 15% respectively. Therefore, it can be concluded that both wind/grid and wind/PV/grid connected systems are feasible in the Mount Hotham both economically and environmentally though the wind has more contribution than the solar energy as wind is available 24 hours a day while solar is available only 7 to 8 hours a day.

Similarly, optimization results for Falls Creek and Mount Buller are analyzed and their performance metrics are compared. **Figure 8** shows the optimization results for Falls Creek and it has been seen that for COE and NPC of wind/grid system are 0.248 $/kWh and $1,878,971 respectively. These values are compared with wind/PV/grid system's performance metrics and it is observed that the system is giving an optimum solution meaning that this location is also suitable for both wind/grid and wind/PV/grid systems.

Optimization results for Mount Buller location with an average wind speed of 5.6 m/sec and with a solar irradiance of 4.33 kWh/m^2 are shown in **Figure 9**. It can be seen that the renewable fraction of 0.71 is observed with a total NPC of $2,208,884 for wind/grid system where as for a wind/PV/grid system observed NPC are $2,254,134 with a RF of 0.74. For PV/grid connected system RF is only 40% with a COE of $0.367. Therefore, from these analyses it can be clearly evident that Mount Hotham is the best suitable location in North East region of Victoria for wind energy generation.

On the other hand, Wangaratta, Edi upper and Benalla are identified as the most suitable places for solar

Figure 6. Optimum results for Mount Hotham location.

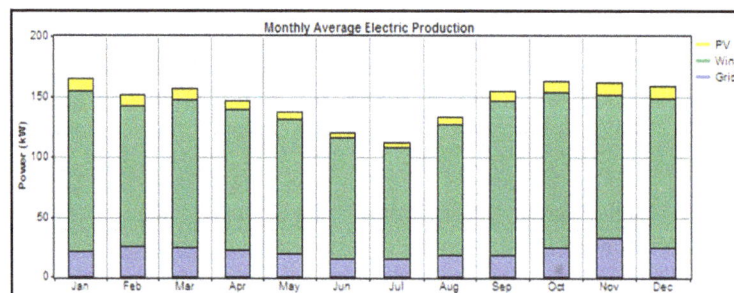

Figure 7. Average electricity production for Mount Hotham.

Figure 8. Optimum results for Falls Creek location.

energy generation. From **Figure 10**, it can be seen that the performance metrics for PV/grid module for Wangaratta are much better than the only grid based system. The NPC and COE of PV/grid connected system are $2,479,722 and 0.343 $/kWh while in only grid connected system are $2,891,005 and 0.400 respectively. Wind energy generation in this location is not promising at all due to unavailability of wind resources.

Figure 11 and **Figure 12** show the simulation results for Edi upper and Benalla locations. In Edi Upper, COE and NPC are 0.356 $/kWh and $2,694,28 while in Benalla, COE and NPC are 0.360 $/kWh and $2,722,924 for a PV/grid connected system respectively. Considering measured performance metrics, PV/grid connected system is more promising compare to PV/wind/grid connected, wind/grid connected and only grid connected system due to high availability of solar resources and poor availability of wind sources in these locations. However, Edi Upper has better prospects of wind energy compare to Wangaratta and Benalla.

Finally, **Table 2** detailed the performance measures of the four studied system. It can be stated that, considering measured performance metrics Mount Hotham is the best location for wind energy generation while Falls Creek and Mount Buller is the second and third suitable location respectively. On the other hand, Wangaratta is

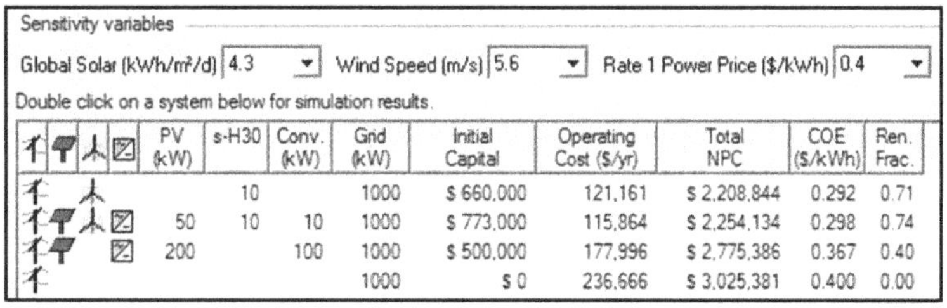

Sensitivity variables
Global Solar (kWh/m²/d) 4.3 Wind Speed (m/s) 5.6 Rate 1 Power Price ($/kWh) 0.4
Double click on a system below for simulation results.

PV (kW)	s-H30	Conv. (kW)	Grid (kW)	Initial Capital	Operating Cost ($/yr)	Total NPC	COE ($/kWh)	Ren. Frac.
	10		1000	$ 660,000	121,161	$ 2,208,844	0.292	0.71
50	10	10	1000	$ 773,000	115,864	$ 2,254,134	0.298	0.74
200		100	1000	$ 500,000	177,996	$ 2,775,386	0.367	0.40
			1000	$ 0	236,666	$ 3,025,381	0.400	0.00

Figure 9. Optimum results for Mount Buller location.

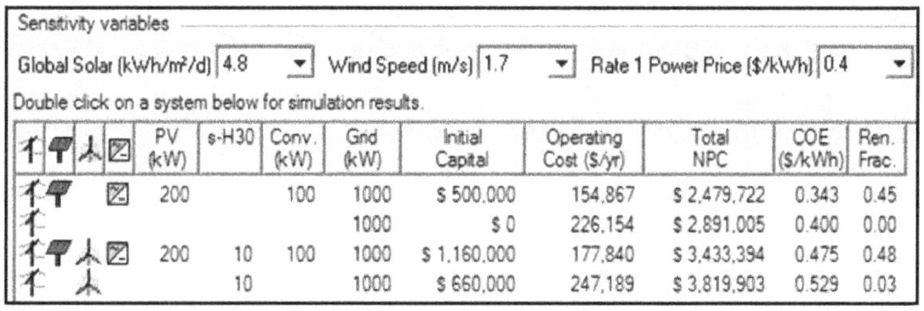

Sensitivity variables
Global Solar (kWh/m²/d) 4.8 Wind Speed (m/s) 1.7 Rate 1 Power Price ($/kWh) 0.4
Double click on a system below for simulation results.

PV (kW)	s-H30	Conv. (kW)	Grid (kW)	Initial Capital	Operating Cost ($/yr)	Total NPC	COE ($/kWh)	Ren. Frac.
200		100	1000	$ 500,000	154,867	$ 2,479,722	0.343	0.45
			1000	$ 0	226,154	$ 2,891,005	0.400	0.00
200	10	100	1000	$ 1,160,000	177,840	$ 3,433,394	0.475	0.48
	10		1000	$ 660,000	247,189	$ 3,819,903	0.529	0.03

Figure 10. Optimum results for Wangaratta location.

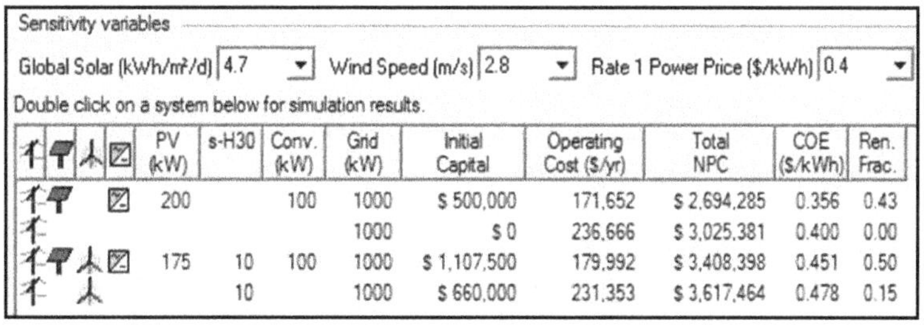

Sensitivity variables
Global Solar (kWh/m²/d) 4.7 Wind Speed (m/s) 2.8 Rate 1 Power Price ($/kWh) 0.4
Double click on a system below for simulation results.

PV (kW)	s-H30	Conv. (kW)	Grid (kW)	Initial Capital	Operating Cost ($/yr)	Total NPC	COE ($/kWh)	Ren. Frac.
200		100	1000	$ 500,000	171,652	$ 2,694,285	0.356	0.43
			1000	$ 0	236,666	$ 3,025,381	0.400	0.00
175	10	100	1000	$ 1,107,500	179,992	$ 3,408,398	0.451	0.50
	10		1000	$ 660,000	231,353	$ 3,617,464	0.478	0.15

Figure 11. Optimum results for Edi upper location.

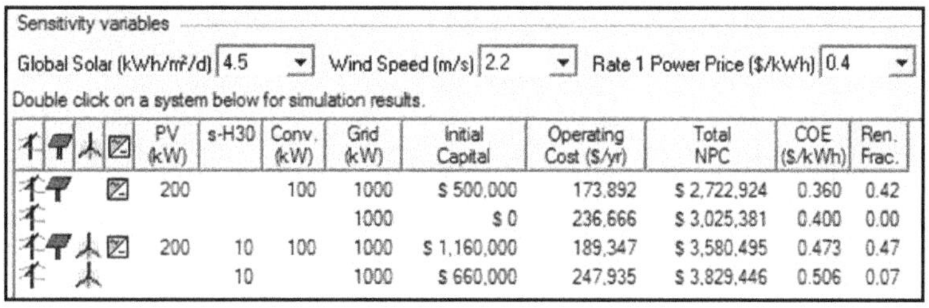

Sensitivity variables
Global Solar (kWh/m²/d) 4.5 Wind Speed (m/s) 2.2 Rate 1 Power Price ($/kWh) 0.4
Double click on a system below for simulation results.

PV (kW)	s-H30	Conv. (kW)	Grid (kW)	Initial Capital	Operating Cost ($/yr)	Total NPC	COE ($/kWh)	Ren. Frac.
200		100	1000	$ 500,000	173,892	$ 2,722,924	0.360	0.42
			1000	$ 0	236,666	$ 3,025,381	0.400	0.00
200	10	100	1000	$ 1,160,000	189,347	$ 3,580,495	0.473	0.47
	10		1000	$ 660,000	247,935	$ 3,829,446	0.506	0.07

Figure 12. Best Optimum results for Benalla location.

Table 2. Performance metrics comparisons for six locations with all possible combinations.

Location	Wind/grid			PV/grid			Wind/PV/grid			Grid only		
	NPC	COE	RF	NPC	COE	RF	NPC	COE	RF	NPC	COE	RF
Mount Hotham	1,649,920	0.218	0.87	3,072,912	0.406	0.15	1,844,260	0.244	0.88	3,025,381	0.400	0.00
Falls Creek	1,878,971	0.248	0.79	2,772,063	0.400	0.40	1,941,005	0.257	0.81	3,025,381	0.400	0.00
Mount Buller	2,208,844	0.292	0.71	2,775,386	0.367	0.40	2,254,134	0.298	0.74	3,025,381	0.400	0.00
Wangaratta	3,819,903	0.529	0.03	2,479,722	0.343	0.45	3,433,394	0.475	0.48	2,891,005	0.400	0.00
Edi Upper	3,617,464	0.478	0.15	2,694,285	0.356	0.43	3,408,398	0.451	0.50	3,025,381	0.400	0.00
Benalla	3,829,446	0.506	0.07	2,722,924	0.360	0.42	3,580,495	0.473	0.47	3,025,381	0.400	0.00

the best place for solar energy generation while Edi Upper and Benalla are the second and third suitable places respectively. Considering combination of both wind and PV generation Mount Hotham is the best place in the North East region of Victoria. However, wind energy has more contribution compare to solar PV in the hybrid power system due to their continuous availability.

3.2. Sensitivity Results

Sensitivity analysis shows the impact of variation in the solar irradiation and the wind speed on the performance of the hybrid system. Based on the scaled annual average value the suitable variables are selected and the simulations are verified for each case. The sensitivity variables considered for solar irradiation are 5.0, 5.5, 6.0, 6.5, 7.0 and for wind speed are 6.9, 7.0, 7.5, 8.0. **Figure 13** shows the line graph between the wind speed and the carbon dioxide emission for Mount Hotham location and from the graph it is clear that the with the increase in the wind speed there is a gradual decrease in the carbon dioxide emission and increase in renewable fraction reducing the GHG emissions in the atmosphere.

From **Figure 14** it has seen that with the increase of solar radiation, total operating cost of energy gener ation is decreasing as well as CO_2 emissions is decreasing as energy generation from solar is increasing.

Figure 15 shows the sensitivity results of Mount Hotham location in which it has seen that a wind/PV/grid connected system is only suitable when the solar radiation is above 4.8 $kWh/m^2/d$ otherwise wind/grid connected system is preferable. This figure also represents the total operating cost with different meteorological conditions and it has observed that the cost of energy generation is decreasing with the increased integration of wind and solar energy into the grid.

Figure 16 shows the surface plot for Wangaratta location in which Grid energy cost and Total Net Present Cost are measured with a fixed wind speed of 1.7m/sec by setting global solar and electricity price as variables. Different possible combinations are verified and it is found that the system would be optimum at certain sensitivity variable resulting in a feasible solution.

4. Conclusions and Future scope

This paper analyses the prospects of Renewable energy in Semi-arid region. A Hybrid model was developed for the North East region of Victoria and identified the best locations for renewable energy generation. A comparison is made among the six locations in terms of NPC, COE, and RF. It is observed that a location which is suitable for wind energy generation also feasible for wind/PV/grid system due to the maximum wind probability. It was also found that Mount Hotham is the most suitable place for wind generation followed by Falls Creek and Mount Buller in the North East region of Victoria. These locations are also suitable for wind/PV/grid systems. On the other hand, solar is feasible in Wangaratta, Edi Upper and Benalla locations in which Wangaratta is the best suitable location. From the optimization and sensitivity results, it is clearly evident that integration of renewable energy, in particular, wind and solar energy into the grid not only reduces the energy crisis worldwide but also reduces the energy costs and GHG emissions. Therefore, renewable energy plays a significant role in developing a climate-friendly sustainable society for the future both nationally and internationally. This fundamental study can be used by the utilities to facilitate large-scale renewable energy integration into the grid. This

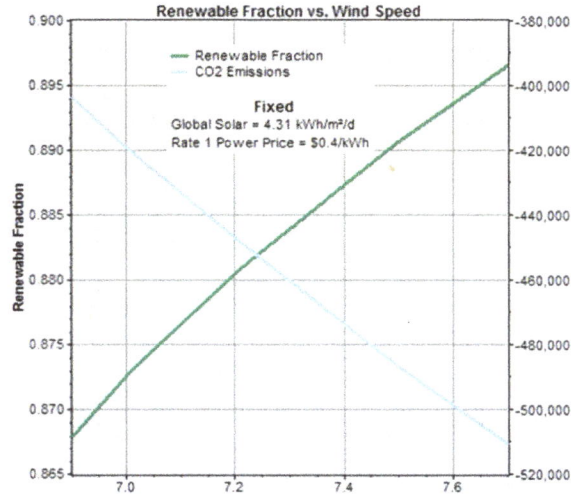

Figure 13. Comparison of wind speed and CO_2 emissions.

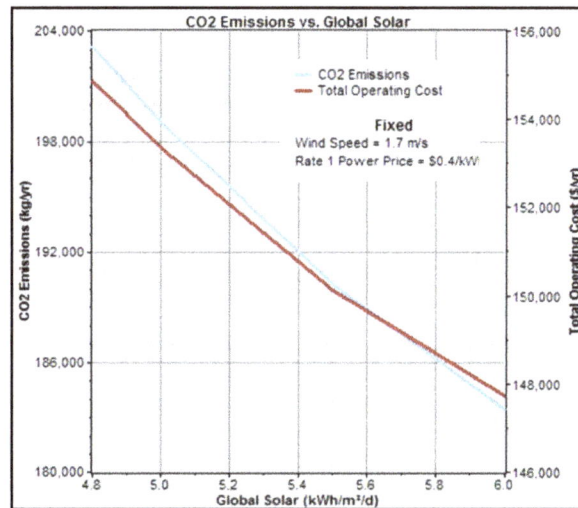

Figure 14. Comparison of solar irradiation and CO_2 emissions.

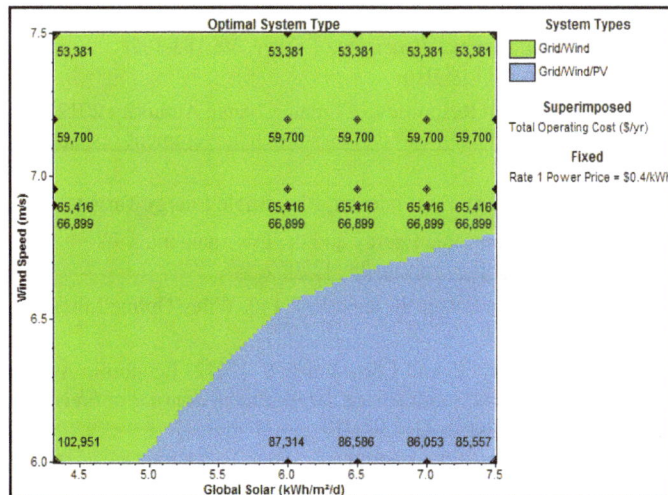

Figure 15. Sensitivity analysis for Mount Hotham location.

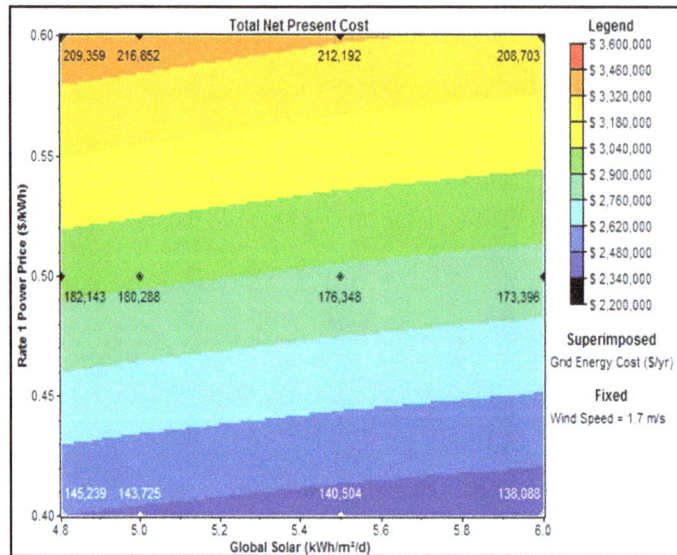

Figure 16. Sensitivity analysis for Wangaratta location.

study is still in its preliminary phase, and further investigations are required in the following areas:

- Details analysis with a larger volume of data considering transmission and other socio-environmental factors;
- Integration of a storage system into the model and compare the revised performance metrics.

References

[1] Abbasi, T. and Sa, A. (2010) Renewable Energy Sources: Their Impact on Global Warming and Pollution. PHI Learning Pvt. Ltd., New Delhi.

[2] Akella, A., Saini, R. and Sharma, M. (2009) Social, Economical and Environmental Impacts of Renewable Energy Systems. *Renewable Energy*, **34**, 390-396. http://dx.doi.org/10.1016/j.renene.2008.05.002

[3] Gol, O. (2008) Renewable Energy—Panacea for Climate Change? *Proceedings of ICREPQ*, **8**, 28.

[4] Shafiullah, G., Amanullah, M., Ali, A.S., Jarvis, D. and Wolfs, P. (2010) Economic Analysis of Hybrid Renewable Model for Subtropical Climate. *International Journal of Thermal & Environmental Engineering*, **1**, 57-65. http://dx.doi.org/10.5383/ijtee.01.02.001

[5] Foster, R., Ghassemi, M. and Cota, A. (2009) Solar Energy: Renewable Energy and the Environment. CRC Press, Boca Raton. http://dx.doi.org/10.1201/9781420075670

[6] Shafiullah, G., Amanullah, M., Shawkat Ali, A., Jarvis, D. and Wolfs, P. (2012) Prospects of Renewable Energy—A Feasibility Study in the Australian Context. *Renewable Energy*, **39**, 183-197. http://dx.doi.org/10.1016/j.renene.2011.08.016

[7] Authority, A.C.C. (2013) Government Response to Climate Change Authority 2012—Review of RET. http://www.climatechange.gov.au/reducing-carbon/news-article/government-response-climate-change-authoritys-2012-review-ret-scheme

[8] A.G.D.O. Environment (2001) C.C.D.O. Environment, Renewable Energy Target.

[9] Nelson, V. (2009) Wind Energy: Renewable Energy and the Environment. http://deakin.eblib.com.au/patron/FullRecord.aspx?p=427069

[10] Ackermann, T. (2005) Wind Power in Power Systems Vol. 140. Wiley Online Library. http://dx.doi.org/10.1002/0470012684

[11] Dekker, J., Nthontho, M., Chowdhury, S. and Chowdhury, S. (2012) Economic Analysis of PV/Diesel Hybrid Power Systems in Different Climatic Zones of South Africa. *International Journal of Electrical Power & Energy Systems*, **40**, 104-112. http://dx.doi.org/10.1016/j.ijepes.2012.02.010

[12] Dalton, G., Lockington, D. and Baldock, T. (2009) Case Study Feasibility Analysis of Renewable Energy Supply Options for Small to Medium-Sized Tourist Accommodations. *Renewable Energy*, **34**, 1134-1144. http://dx.doi.org/10.1016/j.renene.2008.06.018

[13] Beccali, S.B.M., Cellura, M. and Franzitta, V. (2008) Energy, Economic and Environmental Analysis on Ret-Hydrogen Systems in Residential Buildings. *Renewable Energy*, **33**, 366-382.

[14] Liu, G., Rasul, M., Amanullah, M. and Khan, M.M.K. (2011) Feasibility Study of Stand-Alone PV-Wind-Biomass Hybrid Energy System in Australia. *Power and Energy Engineering Conference (APPEEC)*, 1-6.

[15] Liu, M.G.R.G. and Amanullah, M.T.O. (2010) Economic and Environmental Modeling of a Photovoltaic-Wind-Grid Hybrid Power System in Hot Arid Australia.

[16] HOMER (2010) HOMER Energy. http://www.homerenergy.com/

[17] Khan, M. and Iqbal M., (2005) Pre-Feasibility Study of Stand-Alone Hybrid Energy Systems for Applications in Newfoundland. *Renewable energy*, **30**, 835-854. http://dx.doi.org/10.1016/j.renene.2004.09.001

[18] Bureau Australia (2013) Bureau of Metrology. http://www.bom.gov.au/

[19] AEOLOS (2013) AEOLOS Wind Turbine. http://www.windturbinestar.com/

[20] van Alphen, K., van Sark, W.G. and Hekkert, M.P. (2007) Renewable Energy Technologies in the Maldives-Determining the Potential. *Renewable and Sustainable Energy Reviews*, **11**, 1650-1674. http://dx.doi.org/10.1016/j.rser.2006.02.001

21

Australian Consumer Attitudes and Decision Making on Renewable Energy Technology and Its Impact on the Transformation of the Energy Sector

Jeff Sommerfeld, Laurie Buys

Queensland University of Technology, Brisbane, Australia
Email: j.sommerfeld@qut.edu.au, l.buys@qut.edu.au

Abstract

This paper critically examines research on consumer attitudes and behavior towards solar photovoltaic (PV) and renewable energy technology in Australia. The uptake of renewable energy technology by residential consumers in Australia in the past decade has transformed the electricity supply and demand paradigm. Thus, this paper reviews Australian research on consumer behavior, understanding and choices in order to identify gaps in knowledge. As the role of the consumer transforms, there is a critical need to understand the ways that consumers may respond to future energy policies to mitigate unforeseen negative social and economic consequence of programs designed to achieve positive environmental outcomes.

Keywords

Solar Photovoltaic, Renewable Energy, Residential Consumers, Energy Policy, Australia, Transformation, Feed-in-Tariffs

1. Introduction

In Australia, electricity generation is seen as the key for high quality of life and economic development, yet it is responsible for 35 per cent of greenhouse gas (GHG) emissions [1]. For most of the past century, the dominant paradigm of the electricity demand and supply sector has been a technology-push versus consumer demand-pull which has defined traditional market participants [2]. The traditional linear dichotomy of the electricity sector has been rapidly transformed in the past decade, with a demand-pull by residential consumers seeking technological alternatives that supply more environmentally sustainable and cheaper electricity. Residential consumers who were once at the end of the energy supply chain are now using technology to transform themselves into producers and exporters of electricity.

As the role of the residential consumer transforms, there is a need to understand how the consumer will engage future energy policy in an era of technological change. Research into the uptake of energy technology has been identified as being narrowly focused and limited the ability of policy makers to make informed decisions to deal with the complexity of the modern electricity supply paradigm [3]. Recent transformation of the electricity sector has had negative social and economic consequence as a result of the implementation of policies designed to achieve positive environmental outcomes. An examination of accumulated Australian research, with supporting international knowledge on customer behavior to solar and renewable energy technology is timely. The purpose of this article is to explore available Australian research on the energy behavior, understanding and choices of residential consumers in order to identify gaps in knowledge that may guide future research specific to renewable energy and consumer engagement.

2. Background of Australia's Electricity Market

Historically, Australia's small population and vast expanses has been the driver for government initiation of key infrastructure. In late 19th century the electricity industry in Australia was based around numerous small utilities located in individual regional areas with the transfer to State ownership almost complete by the late 1940s. The National Electricity Market (NEM) that exists today did not evolve until the late 20th century [4]. Currently, Australia's electricity network has more than $100 billion in assets with an electricity generation capacity of 54 gigawatts, more than 785,000 km of overhead transmission and distribution lines and more than 124,000 km of underground cables covering vast distance to serve a relatively small population of 23 million people [5].

Global economic and environmental policy shifts in the late 20th century have been the catalyst for change in the Australian electricity sector, including a move towards less government involvement, greater deregulation of energy markets and improved environmental outcomes [4]. The resultant policies led to greater emphasis on renewable energy and the emergence of solar as transformation technology. Government, business and individuals have become increasingly aware of the need to reduce our environmental impact and many individuals have moved beyond mere compliance by engaging in environmental friendly behaviors [6]. Renewable sources of energy are viewed as the most economically viable and environmentally sound options to meet the growing energy needs of the world until technological and safety breakthroughs with other low emission technologies are achieved [7]. This has led to more than 100 countries implementing policies that provide support for renewable power generation and many of these include measures that support domestic solar PV [8]. These economic and environmental policy changes have resulted in major electricity market upheaval and subsequent economic impacts with electricity prices increasing by more than 100 per cent in the past decade and contributing to the demand-pull for technology by consumers seeking alternatives to control costs.

3. Changing Technology

The challenges for electricity markets internationally are significant with socio-economic changes and technological developments posing complex adaptation dilemmas for policy makers and utilities [4]. One of the most significant transformations, since 2001, has resulted from enhanced solar technologies and the domestic rooftop solar PV system. First patented in 1954 [9] the solar cell has in the past decade emerged as a major alternative source of electricity generation. Solar PV systems convert light energy directly into electricity by transferring sunlight photon energy into electrical energy, whereas solar hot water systems use solar radiation to heat water [10] [11]. Consumer demand for solar and renewable energy and resultant government policies and incentives has given rise to almost11 per cent of the Australia population (about 2.6 million people) now using solar for their electricity.

Ongoing technological change such as battery storage is likely to further transform the energy demand and supply paradigm [12]. Transformational technology and the changing role of the consumer has the potential to impact traditional energy market participants which are likely to be faced with lessened demand for electricity from grids. Currently cost of infrastructure and electricity production is shared amongst all consumers. As customers reduce demand or opt out of the electricity supply system due to transformational technology, a diminishing number of customers will directly pay for the electricity supply system. Many of these customers are currently unable to access solar PV and renewable energy technology due to low income (affordability) or living arrangements (renting) [13] [14]. Upward pressure on electricity prices has the potential to migrate greater numbers of consumers to alternatives to the traditional electricity market.

Socially, the move towards alternative energy sources has major ramifications for government policy, given the impact on consumers least able to afford new technologies such as solar PV and batteries. Customers from lower socio-economic demographics often spend a higher proportion of their income on energy and struggle to pay current electricity costs [11] [15]. The structure of incentives for uptake of solar PV technology such as Feed-in-Tariffs (FiTs) is mostly funded from higher electricity charges passed on to all customers [13]. The policies that encourage consumer investment in solar PV and renewable energy technology also have an impact on the electricity network which was designed for a one-way flow of electricity but now must cater to domestic customers feeding solar electricity back into the grid.

The costs of these network upgrades supporting rooftop solar are paid for by all consumers further adding to the costs of people on lower incomes [11] [15]. In Australia and internationally socially regressive aspects of solar, policies have resulted in the transfer of income from lower socio-economic groups to higher socio-economic groups. In many cases only higher socio-economic groups have possessed the necessary access to knowledge and capital that has enabled them to take advantage of solar programs [10] [13] [14] [16].

4. Australian Solar and Renewable Energy Programs

For 20 years, government policies have focused on reducing the cost of solar technologies for consumers and encouraging their uptake. These policies focused on several stages of the energy production chain including rebates for solar water heating systems and residential PV installations [11]. The proportion of households with solar water heaters doubled between 1999 and 2011 [17]. In 2001, the Australian Government introduced the Mandatory Renewable Energy Target (MRET) scheme to encourage investment in renewable energy technologies [17]. The scheme was split in 2010 into two parts: the Large-scale Renewable Energy Target (LRET) and the Small-scale Renewable Energy Scheme (SRES). During this period the Australian Government provided rebates to householders who acquired solar PV systems called the Photovoltaic Rebate Program (PVRP) which was rebranded in 2007 as the Solar Homes and Communities Plan (SHCP) [10] The SRES provided a fixed upfront incentive of about $5000 to reduce the capital cost of solar PV technology while most States and Territories offered the owners of small-scale solar PV installations a Feed-in-Tariff (FiT) that paid households for electricity generated [16].

As a result of consumer demand and resultant government policies and incentives over one million rooftop solar PV systems have been installed in Australia. In just four years, between 2007 and 2011, the cumulative installed capacity of solar PV units increased 100-fold from about 10 MW to more than 1000 MW [16]. The state of Queensland, known as the "Sunshine State", has the largest number of solar PV installations of any state, followed by New South Wales and Victoria [18].

5. Consumer Uptake of Solar and Renewable Energy Technology

For most of the past century the dominant paradigm of the electricity demand and supply sector has been a technology-push versus consumer demand-pull for this technology which has defined traditional market participants [2]. In recent years, better informed consumers are increasingly taking into consideration the environmental and social impact of products and services [19]. As a consequence of consumer demand and resulting government policy there has been a demand-pull for better environmental, economic and social sustainability that has transformed the traditional linear dichotomy of the electricity sector. Consumers who were once at the end of the energy supply chain are now using technology to transform themselves into producers and exporters of electricity. Yet, much of the research to date has focused on either the reasons for adoption or non-adoption of renewable energy or the social consequences from it.

Prior to the surge in uptake in solar PV from 2008, Caird and colleagues [20] undertook a study of consumers surveying reasons for adoption or non-adoption of renewable energy and energy efficiency measures. This research drew together previous quantitative surveys of consumers from the UK, USA and Australia on attitudes to renewable energy and installation barriers. The main drivers for installation were environmental concern and saving money whilst the main barriers were capital cost and lack of trustworthy information or reliable brands. It was concluded that research tended to focus on addressing financial, regulatory and information barriers and drivers. The researchers identified that social context was crucial in understanding consumer energy behavior and sociological and anthropological research focusing on motivations and actions suggested consumer motivations were more complex.

Research focusing on the financial uptake of solar PV and renewable energy, examined the impact of policy mechanism used to encourage consumer uptake of solar PV such as the solar FiT [10] [13] [14] [16]. An evaluation of the Australian Government Photovoltaic Rebate Program (PVRP), later rebranded the Solar Homes and Communities Plan, concluded the program was environmentally ineffective, economically costly and had social equity issues [10]. This Australian finding is similar to an examination of German climate policy that encouraged the uptake of solar PV and use of FiTs [13]. In the decade between 2000 and 2011 the share of renewables in Germany increased from seven to 20 per cent. Whilst FiT policies in Germany encouraged the dissemination of renewables technology, subsidies that unpinned the expansion increased from 900 million Euros to 16.7 billion Euros was funded by adding three Euro cents per kilowatt hour to the cost of bills. The authors concluded these policies were regressive as they facilitated that expansion of expensive technology without fostering cost-reducing innovation and had a negligible impact on climate protect [13].

Social context to customer decisions to adopt or non-adopt solar PV and renewable energy technology appears to have so far attracted limited research interest to date. Hampton and Eckermann [21] explored the ways social learning can be used to improve understanding of solar PV's based on changing attitudes. Through qualitative workshops in 2005 and 2012, knowledge and understanding of solar PV and renewable energy products was found to have considerably improved during the two workshops but customers still had difficulties understanding financial aspects of solar PV policy.

The profile of consumers adopting renewable energy technologies appears to be inconclusive with investigations into educational status and environmental behavior providing conflicting evidence. Demographic variables associated with positive environmental attitudes such as age, gender and income have identified conflicting conclusions [3]. For example, the researchers identified studies that found having a higher education level encouraged environmentally positive behavior whilst another study found less educated consumers were more likely to be green consumers. To overcome demographic variations a study of Australian consumers cross referenced both socio-economic status based on income and value of the housing [22]. They concluded that lower income households were not engaging solar PV and renewable energy technology and owning a property was found to be the most important criteria in decisions to adopt or non-adopt solar PV and renewable energy technology.

6. Impacts of Consumer Uptake of Solar and Renewable Energy Technology

Whilst consumers may have been mentioned in most research, the majority of researchers focused on energy policy at the national and international level with specific attention on government energy policies and the implications of these policies. Solar PV has diverse economic, environmental and social values and policies encouraging it has generally been developed and implemented without any comprehensive social cost-benefit analysis being undertaken [8]. The research examining consumer uptake of energy technology focused on addressing financial, regulatory and information barriers and drivers. Underpinning much of the research is a primary assumption that environmental outcomes are the key indicator of success. This type of examination looked at societal values from solar PV including carbon abatement, consumer outcomes from deferring network augmentation and offsetting energy losses [8] [23]-[25]. Evaluations of the policies that encourage the uptake of renewable technology found that only some home owners had the capacity to afford and install solar PV systems based on socio-economic profiles [10] [13] [14] [16]. In examining the uptake of solar energy policies, the type of housing was identified as an obstacle (e.g. apartment, unit) or living arrangements (e.g. renting) [14]. Overall the policies that encouraged renewables also were found to impose additional networks costs that were funded by consumers not using renewables. These customers were further disadvantaged if they were on lower incomes as they spend a higher proportion of their income on energy [11] [15].

In an examination of government policies that encouraged the technological transformation, Taylor [2] concluded the effectiveness of innovation was not a primary consideration. Immediate pollution reduction and energy conservation has been the policy drivers rather than an empirical evaluation of the comparative effects of various options. Additionally, research into the uptake of solar and renewable energy technology by consumers is mostly silent on the impact of renewable energy uptake on other consumers. The unforeseen outcome of consumer decisions to adopt or not adopt solar and renewable energy technology has been an increase in the social divide between consumers [10] [13]-[15].

7. Conclusions

Currently the research focus is on single aspects of environmental, economic or social attitudes of consumers and the impact on the electricity sector or electricity policy [11] [19] [26]. Other researchers tended to focus on policy, policy-induced technical change, financial issues and consumer environmental attitudes [6] [9] [27]. Policy and policy implications were explored, but the investigations did not extend to examining the consumer behavior resulting from these policies. Whilst the phenomena relating to consumer energy was a key research focus, researchers did not explain the motivation or context of consumers who adopted or did not adopt renewables technology. As a result, much of the research is inconclusive with regard to understanding consumer behaviour.

Consumers were examined from a macro perspective rather than the more complex approach recommended by Caird *et al.* [20] and Faiers *et al.* [3]. Conclusions on the effectiveness of solar and renewable energy technology policy need to address the complex social, economic and environmental interactions and outcomes that lead to a holistic understanding and insight into the complexity of energy use and impact. Research needs to address this complexity in order to identify and integrate the social, technical and environmental changes and their impact on the diverse groups of consumers.

Internationally, a review of the European *Residential Monitoring to Decrease Energy Use and Carbon Emissions in Europe* (REMODECE) project identified the importance of ongoing research to track the influence of new trends in technology and consumer behavior. The REMODECE project was established to better understand household energy consumption and identify demand trends. It concluded that research examining consumer uptake of energy technology must encompass personal values and attitudes and the impact of external factors [28]. Research needs to go beyond cognitive assessment and rational choice because of emotional, societal and cultural issues impact on consumer energy behavior [3].

In conclusion, the purpose of this paper was to examine contemporary research on consumer behavior, understanding and choices towards solar technology in Australia. With almost 11 per cent of the Australian population now using solar for their electricity, research in this area is essential to developing future policy. The rapid uptake of technology by consumers has not only transformed the demand and supply dichotomy but also social and economic aspects of the electricity market. The consumer decision to acquire a solar PV system is complex requiring information that most average consumers are unlikely to have in early stages of new technology [29]. Research into the consumer uptake of energy technology has been narrowly focused and limits the ability of policy makers to make informed decisions [20]. Understanding the demand-pull social phenomena has significant relevance given the equity issues for low-income consumers. Whilst the adoption of solar PV is positive in terms of environmental concerns, researchers have failed to adequately examine the resultant economic or social consequences across user groups. As the role of the consumer transforms, there is a need to understand how the consumer will engage future energy policy to mitigate unforeseen negative social and economic consequence of programs designed to achieve positive environmental outcomes.

References

[1]   Evans, A., Strezov, V. and Evans, T.J. (2010) Sustainability Considerations for Electricity Generation from Biomass. *Renewable and Sustainable Energy Reviews*, **14**, 1419-1427. http://dx.doi.org/10.1016/j.rser.2010.01.010

[2]   Taylor, M. (2008) Beyond Technology-Push and Demand-Pull: Lessons from California's Solar Policy. *Energy Economics*, **30**, 2829-2854. http://dx.doi.org/10.1016/j.eneco.2008.06.004

[3]   Faiers, A., Cook, M. and Neame, C. (2007) Towards a Contemporary Approach for Understanding Consumer Behaviour in the Context of Domestic Energy Use. *Energy Policy*, **35**, 4381-4390.
http://dx.doi.org/10.1016/j.enpol.2007.01.003

[4]   Quezada, G., Grozev, G., Seo, S. and Wang, C.H. (2014) The Challenge of Adapting Centralised Electricity Systems: Peak Demand and Maladaptation in South East Queensland, Australia. *Regional Environmental Change*, **14**, 463-472.
http://link.springer.com/article/10.1007/s10113-013-0480-0

[5]   Kuwahata, R. and Monroy, C.R. (2011) Market Stimulation of Renewable-Based Power Generation in Australia. *Renewable and Sustainable Energy Reviews*, **15**, 534-543. http://dx.doi.org/10.1016/j.rser.2010.08.020

[6]   Gadenne, D., Sharma, B., Kerr, D. and Smith, T. (2011) The Influence of Consumers' Environmental Beliefs and Atti-

tudes on Energy Saving Behaviours. *Energy Policy*, **39**, 7684-7694. http://dx.doi.org/10.1016/j.enpol.2011.09.002

[7]   Sener, C. and Fthenakis, V. (2014) Energy Policy and Financing Options to Achieve Solar Energy Grid Penetration targets: Accounting for External Costs. *Renewable and Sustainable Energy Reviews*, **32**, 854-868. http://dx.doi.org/10.1016/j.rser.2014.01.030

[8]   Sebastián Oliva H., Mac Gill, I. and Passey, R. (2014) Estimating the Net Societal Value of Distributed Household PV Systems. *Solar Energy*, **100**, 9-22. http://dx.doi.org/10.1016/j.solener.2013.11.027

[9]   Peters, M., Schneider, M., Griesshaber, T. and Hoffmann, V.H. (2012) The Impact of Technology-Push and Demand-Pull Policies on Technical Change: Does the Locus of Policies Matter? *Research Policy*, **41**, 1296-1308. http://dx.doi.org/10.1016/j.respol.2012.02.004

[10]   Macintosh. A. and Wilkinson, D. (2011) Searching for Public Benefits in Solar Subsidies: A Case Study on the Australian Government's Residential Photovoltaic Rebate Program. *Energy Policy*, **39**, 3199-3209. http://dx.doi.org/10.1016/j.enpol.2011.03.007

[11]   Bahadori, A., Nwaoha, C., Zendehboudi, S. and Zahedi, G. (2013) An Overview of Renewable Energy Potential and Utilisation in Australia. *Renewable and Sustainable Energy Reviews*, **21**, 582-589. http://dx.doi.org/10.1016/j.rser.2013.01.004

[12]   Rudolf, V. and Papastergiou, K.D. (2013) Financial Analysis of Utility Scale Photovoltaic Plants with Battery Energy Storage. *Energy Policy*, **63**, 139-146. http://dx.doi.org/10.1016/j.enpol.2013.08.025

[13]   Grösche, P. and Schröder, C. (2011) On the Redistributive Effects of Germany's Feed-in-Tariff. *Empirical Economics*, **46**, 1339-1383. http://dx.doi.org/10.1007/s00181-013-0728-z

[14]   Byrnes, L., Brown, C., Foster, J. and Wagner, L.D. (2013) Australian Renewable Energy Policy: Barriers and Challenges. *Renewable Energy*, **60**, 711-721. http://dx.doi.org/10.1016/j.renene.2013.06.024

[15]   Bell, W.P. and Foster, J. (2013) Feed-in-Tariffs for Promoting Solar PV, Energy Storage and Other Distributed Resources: Progressing from Calculated to Market Determined Feed-in-Tariffs: Part 1 and 2. MPRA Paper. http://mpra.ub.uni-muenchen.de/id/eprint/49527

[16]   Nelson, T., Simshauser, P. and Nelson, J. (2012) Queensland Solar Feed-in-Tariffs and the Merit-Order Effect: Economic Benefit, or Regressive Taxation and Wealth Transfers. *Economic Analysis and Policy*, **42**, 277-301.

[17]   Ferrari, D., Guthrie, K., Ott, S. and Thomson, R. (2012) Learning from Interventions Aimed at Mainstreaming Solar Hot Water in the Australian Market. *Energy Procedia*, **30**, 1401-1410. http://dx.doi.org/10.1016/j.egypro.2012.11.154

[18]   Flannery, T.F. and Sahajwalla, V. (2013) The Critical Decade: Australia's Future: Solar Energy. Climate Commission Secretariat, Department of Industry, Innovation, Climate Change, Science, Research and Tertiary Education, Canberra.

[19]   Auger, P., Devinney, T.M., Louviere, J.J. and Burke, P.F. (2010) The Importance of Social Product Attributes in Consumer Purchasing Decisions: A Multi-Country Comparative Study. *International Business Review*, **19**, 140-159. http://dx.doi.org/10.1016/j.ibusrev.2009.10.002

[20]   Caird, S., Robin, R. and Herring, H. (2008) Improving the Energy Performance of UK Households: Results from Surveys of Consumer Adoption and Use of Low- and Zero-Carbon Technologies. *Energy Efficiency*, **1**, 149-166. http://dx.doi.org/10.1007/s12053-008-9013-y

[21]   Hampton, G. and Eckermann, S. (2013) The Promotion of Domestic Grid-Connected Photovoltaic Electricity Production through Social Learning. *Energy, Sustainability and Society*, **3**, 23. http://link.springer.com/article/10.1186/2192-0567-3-23 http://dx.doi.org/10.1186/2192-0567-3-23

[22]   Nelson, T., Simshauser, P. and Kelley, S. (2011) Australian Residential Solar Feed-in-Tariffs: Industry Stimulus or Regressive Form of Taxation. *Economic Analysis and Policy*, **41**, 113-129.

[23]   Solangi, K.H., Islam, M.R., Saidur, R., Rahim, N.A. and Fayaz, H. (2011) A Review on Global Solar Energy Policy. *Renewable and Sustainable Energy Reviews*, **15**, 2149-2163. http://dx.doi.org/10.1016/j.rser.2011.01.007

[24]   Timilsina, G.R., Kurdgelashvili, L. and Narbel, P.A. (2012) Solar Energy: Markets, Economics and Policies. *Renewable and Sustainable Energy Reviews*, **16**, 449-465. http://dx.doi.org/10.1016/j.rser.2011.08.009

[25]   Zahedi, A. (2010) A Review on Feed-in-Tariff in Australia, What It Is Now and What It Should Be. *Renewable and Sustainable Energy Reviews*, **14**, 3252-3255. http://dx.doi.org/10.1016/j.rser.2010.07.033

[26]   Martin, N.J. and Rice, J.L. (2012) Developing Renewable Energy Supply in Queensland, Australia: A Study of the Barriers, Targets, Policies and Actions. *Renewable Energy*, **44**, 119-127. http://dx.doi.org/10.1016/j.renene.2012.01.006

[27]   Negro, S.O., Alkemade, F. and Hekkert, M.P. (2012) Why Does Renewable Energy Diffuse So Slowly? A Review of Innovation System Problems. *Renewable and Sustainable Energy Reviews*, **16**, 3836-3846. http://dx.doi.org/10.1016/j.rser.2012.03.043

[28] de Almeida, A., Fonseca, P., Schlomann, B. and Feilberg, N. (2011) Characterization of the Household Electricity Consumption in the EU, Potential Energy Savings and Specific Policy Recommendations. *Energy & Buildings*, **43**, 1884-1894. http://dx.doi.org/10.1016/j.enbuild.2011.03.027

[29] Guidolin, M. and Mortarino, C. (2009) Cross-Country Diffusion of Photovoltaic Systems: Modelling Choices and Forecasts for National Adoption Patterns. *Technological Forecasting & Social Change*, **77**, 279-296. http://dx.doi.org/10.1016/j.techfore.2009.07.003

22

A Methodology for Introducing M&V Adjustments during an Energy Retrofit Impact Assessment

Nikos Sakkas, Evangelos Kaltsis

Applied Industrial Technologies Ltd., Gerakas, Greece
Email: sakkas@apintech.com

Abstract

The assessment of an energy retrofit necessarily requires an energy measurement campaign before (base year energy consumption) and after (post retrofit energy consumption) the retrofit. Only in this way is it possible to reach a safe conclusion, on the true retrofit impact. In addition, a number of adjustments are necessary to secure that the retrofit impact on energy consumption is effectively isolated, *i.e.*, which we report on the true retrofit impact and not, for example, on external variations, such as a more mild winter. This paper introduces a conceptual framework for taking account, in the retrofit impact assessment, of three external parameters: weather, indoor comfort and space occupancy. The broader strategy behind this work is to develop a comprehensive methodology that would allow a cost efficient, fast and accurate assessment of energy retrofits in buildings. This would allow insight, on the investor side, as to the prudence of his investment and, and in this way, could help the proliferation of the practice of energy retrofits. The adjustment methodology, introduced here, is a first step in this direction.

Keywords

Energy Retrofit, Impact Assessment, External Parameter Adjustment, Occupancy, Operational Retrofit

1. Introduction

Building retrofits are an important activity in the construction industry. They may be initiated for many reasons, one of which is the reduction of energy consumption (energy retrofits). What remains a key barrier for a wider uptake of energy retrofits is the difficulty one encounters to measure and communicate their impact. As a result, the lack of data backed evidence on the retrofit impact does not allow the building owner to gain insight in his investment. Similarly, it does not allow the retrofit provider to fully understand his solution limitations and launch continuous improvement strategies.

According to IPMVP (International Performance Measurement and Verification Protocol), measuring energy savings should be based on the following general equation:

$$\text{Energy Savings} = \text{Baseline (pre retrofit) Energy Use} - \text{Post Retrofit Energy Use} \pm \text{Adjustments} \qquad (1)$$

Adjustments are required to account for changes of external conditions. Weather and occupancy information are key external parameters, suggested for use in M&V protocols [1] [2]. They will also be used in our approach. The uncertainty of such data has received significant attention in the literature [3]-[6]. Weather data are often taken from distant meteorological stations and might fail, in this way, to capture micro-climate conditions around the specific building. Occupancy presents important difficulties to measure. Several approaches have been proposed in the literature for assessing baseline as well as post retrofit energy use. Kissock [7] developed a regression methodology to measure retrofitting energy use in commercial buildings. Krarti *et al.* [8] utilized neural networks to estimate energy and demand savings from retrofits of commercial buildings. Dhar *et al.* [9] generalized the Fourier series approach to model hourly energy use in commercial buildings. In addition, in most practical cases, utility bill data are used because they are widely available and inexpensive to obtain and process. Reddy *et al.* [10] presented a formal baselining methodology at the whole building level based on monthly utility bills and took outdoor dry-bulb temperature as the only model regressor.

This research background reveals that adjustments are only one of the many difficulties that need to be overcome towards our end goal, *i.e.*, an accurate, reliable and cost efficient assessment of the retrofit impact. However, even defining the baseline energy consumption, in a cost effective and practical way, may, in some cases, present significant difficulties. Let us imagine, for example, the case where a sub-part (e.g. a floor) of a building, equipped with a centralized HVAC system, is retrofitted. How can one define here the baseline consumption? This is not a trivial question if we wish to end up with a practical solution and avoid costly measurements of the many parameters contributing to it (e.g. air duct velocities). Obviously, scaling down building level data to our reduced retrofit scope (e.g. floor) would introduce large errors and can not be recommended.

Another critical issue for the viability of a retrofit assessment procedure is the duration of the campaign. Any scheme requiring more than a couple of weeks of monitoring before and after the retrofit would stand little chances for business uptake.

In this paper, we touch upon only one of the many issues that are related to this key issue of retrofit assessment. We discuss the issue of M&V adjustments of parameters that may be different before and after the retrofit and may, because of this, affect our assessment. We propose a methodological framework to address three types of possible parameter variation: weather, indoor conditions and building occupancy. This is summarized in **Figure 1**.

2. Methods

We will define our baseline impact, prior to any adjustment, by means of Equation (2). This equation essentially calculates the change of energy consumption, of our study building space, before and after the retrofit, as a percentage of the former energy consumption. Energy consumed is measured in the real time, at an equal, before and after the retrofit, number of time instances [N].

$$\text{Baseline Impact} = \left[\left[\sum_{i=1}^{N} E_{[\text{pre},i]} \right] - \left[\sum_{i=1}^{N} E_{[\text{post},i]} \right] \right] \bigg/ \left[\sum_{i=1}^{N} E_{[\text{pre},i]} \right] \qquad (2)$$

Figure 1. Weather, building occupancy and indoor conditions included as the M&V adjustments to assess the true energy retrofit impact.

pre, the pre-retrofit period, year or other.

post, the post-retrofit period, of a similar duration with the pre-retrofit one.

i, a measurement instance.

N, the number of measurements that we assume will be the same in both, pre and post periods. With regard to the data resolution, the methodology will be able to operate on any resolution. Typically, the "hour" would be a reasonable time scale for data collection.

$E_{[pre,i]}$, the energy consumption at a measurement time [i] of the [pre] period.

$E_{[post,i]}$, the energy consumption at a measurement time [i] of the [post] period.

$\sum_{i=1}^{N} E_{[pre,i]}$, total energy consumption in the pre-retrofit period.

$\sum_{i=1}^{N} E_{[post,i]}$, total energy consumption in the post-retrofit period.

We will now define methods for adjusting this baseline figure for the three external parameters suggested above.

2.1. Adjusting for Outdoor Temperature

Let us consider that $T_{pre,i}$ and $T_{post,i}$ represent the external temperature at time [i] within the [pre] and [post] periods respectively. For simplicity and with no loss of generality the resolution of the temperature data is set similar to that of energy data. In the unlikely case that $T_{pre,i} = T_{post,i}$, for all [i], external temperature will not affect our impact figure in Equation (2). Thus, if the energy consumption before the retrofit is 100 and the respective energy consumption after the retrofit is 80, we can safely assume that the impact of our retrofit is as shown below.

$$\text{Impact} = [100 - 80]/100 = \text{an } 20\% \text{ Reduction/Saving of Energy} \qquad (3)$$

Obviously, if the equation $T_{[pre]} = T_{[post]}$ for all [i] does not apply, the above impact figure is not valid any more. The energy consumption is now affected by the different weather among the two periods.

We propose to adjust for the weather by extending a well known concept in the related literature, that of heating/cooling degree hours. Fels *et al.* [11] were the first to utilize a similar concept, of the variable-base degree-day in retrofitting. A similar approach was used in [12]. Here, we expand this concept by increasing its resolution at the hour level. Heating/cooling degree hours, noted as **Hdh** and **Hch** respectively are a means to describe the thermal needs of a building. Their calculation is very simple and is based on the comparison of the hourly outdoor temperature with a preset reference temperature (e.g. 20 deg). Let us see how this would work.

Let us assume two, equal in time, periods across the year, the first representing a heating period with a reference temperature set to 20 and the second a cooling period with a reference temperature of 26 degrees. A year has 8760 hours, therefore 4380 would be considered as heating hours and an equal 4380 as cooling hours. Any other separation between heating and cooling hours is possible, without any loss of generality. Even three or four periods can be easily accommodated in the methodology.

For every hour of our temperature monitoring during the heating period we would subtract the measured temperature from the reference temperature and calculate the degree hours during that, particular, hour. Following the same rule, across the whole heating period, we would end up with a total of heating degree hours. Expanding over to the cooling period we would calculate, in an identical way, the total of the cooling degree hours. Obviously, the harsher the winter, the higher the heating degree hours; the hotter the summer, the higher the cooling degree hours; and vice versa.

In this way, we have captured in a single metric the impact of weather. Obviously, this is a linear approximation and, because of this, introduces some error. A linear approximation essentially implies that the heating energy required to raise, as an example, the indoor temperature one degree when the outdoor temperature is 8 degrees is the same as that required to raise the indoor temperature one degree when the outdoor temperature is 18 degrees. This, in theory is not true, however in practice it is *close* to being true. The linear approximation proposed is a workable methodology, easy to implement. Any more sophisticated approach would introduce complex phenomena that could perhaps increase the accuracy to, however, a large and unacceptable loss of practicality.

Let us know see an example to see how our impact is adjusted against the outdoor temperature. We will use the same as above example, where the energy consumption before the retrofit is 100 and the respective energy consumption after the retrofit is 80. However, now the equation $T_{pre,i} = T_{post,i}$ for all [i], does not apply any more. In addition, we have carried out our real time calculation of heating degree hours for both periods, before and after the retrofit. Thus, let us say the $Hdh_{pre} = 1500$ and the $Hdh_{post} = 2000$.

The figures above indicate that in the period after the retrofit, the weather has been much harsher as reflected in the significantly higher number of heating degree hours. Our linear model would thus require an offset factor as follows:

$$\text{Weather Adjustment Factor} = Hdh_{post}/Hdh_{pre} = 2000/1500 = 4/3 \qquad (4)$$

Therefore our impact figure, adjusted for weather, denoted by **impact [w]**, would now assume the following value:

$$\text{Impact [w]} = \text{Baseline Impact Weather Adjusement Factor} = 20\%\ 4/3 = 26.7\% \qquad (5)$$

One may now ask whether we need to carry out a similar exercise during the cooling period. A closer look to the retrofit is required to answer this point. If, for example, the retrofit includes elements (e.g. window changes) that affect their g value, monitoring during the cooling period would be necessary as changes of g values will manifest more strongly during the summer period.

2.2. Adjusting for Indoor Comfort

This purpose of this adjustment is to offset any different indoor comfort conditions between the [pre] and [post] periods and to dissociate them from the true retrofit impact. Different indoor conditions may, for example, result because of a different thermostat setting, between the two periods. If users, for any reason, opt for a different indoor temperature before or after the retrofit, this will affect energy consumption figures. To isolate the true retrofit impact one would need to carry out an adjustment, along a similar, as above, rationale. Instead of "heating degree hours" we will now use the concept of "comfort degree hours", noted as **Cdh**. Comfort degree hours are calculated by subtracting the indoor temperature from a reference comfort temperature set, for example at 22 degrees. The calculation is done is an exactly similar way. Real time indoor, now, temperatures are subtracted every [i] time from this reference temperature to calculate the comfort hours. We would again opt for a linear model, for the same reasons as in the case of weather. Also, we restrict to the temperature aspect of indoor climate and do not consider humidity changes. However, an enthalpy, instead of a temperature, adjustment would be perfectly possible to include also humidity changes in the adjustment factor. That would not increase at all the cost or the complexity of the exercise, especially as humidity sensors are of a low cost and often come together with ambient temperature ones.

As an example, let us assume we have calculated a figure of $Cdh_{pre} = 630$ and the $Cdh_{post} = 420$. This would imply that in the post retrofit period the indoor environment was more "cold" resulting to a much decreased value of the comfort degree hours. We would therefore need to compensate this fact and offset it from our impact calculation. Similar as above, we would establish a comfort adjustment factor as follows:

$$\text{Comfort Adjustment Factor} = Cdh_{post}/Cdh_{pre} = 420/630 = 2/3 \qquad (6)$$

One can notice that, in this case, the impact of the different comfort conditions has a negative impact on true impact (factor 2/3). Our new impact indicator, adjusted also for indoor comfort, denoted by **impact [w, c]**, would now be:

$$\text{Impact [w, c]} = \text{Impact [w] Comfort Adjusement Factor} = 26.7\%\ 2/3 = 17.8\% \qquad (7)$$

2.3. Adjusting for Occupancy

Occupant densities in office buildings can vary between 4.3 m^2 and 22.8 m^2 per person, and this range obviously significantly impacts internal heat gains [13] [14]. Overall, it is now widely acknowledged in the literature [15]-[25] that occupant behavior plays a major role in determining building energy use. It is usually the main reason causing the significant gaps between actual and predicted energy performance [16] [26] [27] of buildings. Studies have shown that occupant behavior may vary to such an extent that the resultant building energy use may differ by a factor of two or more! [28] [29].

Building occupancy profiles may change for a number of reasons. For example, the working hours may change. The spaces may change, may increase or temporarily decrease due to maintenance works. Then the number of people may increase, decrease or change across time and space. Unpredictable events, natural catastrophes, strikes, may further add to the occupancy changes.

Do occupancy changes affect our energy considerations? The case deserves here some special attention, as it is not as straightforward as in the above cases, for indoor and outdoor conditions.

Let us imagine an office that accommodates ten office employees and in the post retrofit period their number has increased to twelve. Obviously the energy efficiency (energy consumed per person) of the space has decreased but the energy requirements of it will have remained more or less the same. In this case it would not really make sense to compensate for the new occupancy. On the other hand if, as a result of the retrofit itself, new spaces are made operational or old ones are decommissioned then this occupancy change would have an impact also on the energy profiles. Thus, there are two broad cases for occupancy.

Occupancy changes during retrofits may be due to the change of the effective building **area** or the **timing** of its use. In a way, our reference building after the retrofit has now somehow changed; this will result to a different energy consumption profile. Thus, there is a need to account for the different occupancy before and after the retrofit.

Occupancy changes may also be due to changing space **use profiles**, as in the example above where people office density increased. In this case, it does not seem appropriate to compensate for occupancy, as the energy consumption will only marginally be affected by such changes.

However, in the case of the occupancy, it might not be the best possible approach to keep on looking into the retrofit impact in terms of changes of the energy consumption per year (KWH/year). Indicators that consider **space efficiency rather than consumption (KWH/person/year)** seem much better positioned to model the true occupancy retrofit impact. Indeed, such indicators would now be affected by our "density" like, occupancy, changes.

This discussion highlights a different type of retrofitting; **operational retrofitting**, *i.e.*, changing our buildings' use profiles so that they may serve more people with the same energy, so that they may be more use efficient. At the end, **"energy per serviced user-hours"** would be the ultimate building energy efficiency indicator. It would be far more comprehensive than the usual "energy per hypothetical population" or "energy per footage".

As a summary, we have highlighted above that the issue of occupancy holds an important potential. However, its treatment is not as straightforward as its heating and comfort degree hours peers. The analysis will be different depending on whether we are looking at the retrofit impact on energy consumption or if we are also interested in the more subtle space performance, in the case where our retrofit has included aspects of operational changes or retrofits.

Let us consider these two cases independently.

The classical retrofit and the case of building/space energy demand.

In this case, we assume our retrofit has had by design no intention to affect directly or indirectly the operational profiles of the space/ building. Here we need to investigate, one by one, the occupancy changes that may have resulted. Typical issues to consider here are:

Are new building parts put in operation or are older ones decommissioned? Has their use timing changed?

Such issues will have an impact on the energy consumption. An occupancy based adjustment will be required.

The operational retrofit and the case of building/space energy performance.

In this case, we assume our retrofit has had by design an intention to affect directly or indirectly the operational profiles of the space/building. We are now entitled to have these, now increased, occupancies, incorporated in our impact indicator in a favorable way.

Our adjustment approach will be similar to the two above discussed adjustments, for external and internal temperature. The adjusting parameter will now be the average, headcount, occupancy, denoted in the following as Occ.

Let us assume we have calculated a figure of $Occ_{pre} = 30$ and the $Occ_{post} = 40$. This would imply that in the post retrofit period more people are serviced, either because of the retrofit design or for any other reason as those discussed above (e.g., an increase or the operational building hours). Again, we would now need to compensate this fact and offset it from our impact calculation. Our occupancy adjustment factor would now be calculated as follows:

$$\text{Occupancy Adjustment Factor} = \text{Occ}_{\text{post}}/\text{Occ}_{\text{pre}} = 40/30 = 4/3 \tag{8}$$

Our new impact indicator, adjusted also for occupancy and denoted by impact [w, c, o], would now be,

$$\text{Impact } [w, c, o] = \text{Impact } [w, c] \text{ Occupancy Adjustement Factor} = 17.8\% \, 4/3 = 23.7\% \tag{9}$$

3. Conclusions

We have presented above a conceptual framework for managing and carrying out adjustments when trying to assess the impact of a retrofit. We have suggested that these adjustments should include compensation for different outdoor (change of outdoor temperature), different indoor (change of indoor temperature), changes in building occupancy and use profiles.

Besides emphasizing the need to carry out adjustments along the three above parameters, we have developed a concrete methodology for carrying them out. In the case of indoor and outdoor conditions, the methodology is an extension of the well-known concept of heating days/hours. Heating/cooling hours will offset the weather impact while comfort hours will adjust the indoor climate variation. In this paper, we have modeled the comfort hours exclusively on indoor temperature; this can be readily extended to an enthalpy based indicator, including also the humidity changes.

In the case of occupancy, we have considered it necessary to differentiate between occupancy changes that come as a result of the retrofit itself and those resulting in an *ad hoc* manner. In the first case we have a so called operational retrofit; this should be always accounted for in the retrofit impact calculation. An operational retrofit reconsiders the space use and space user deployment in view of a better energy performance. Operational retrofits hold some important savings' potential and have been somehow overlooked as to their potential to deliver energy savings. However, the way that a building is operated can have an important impact, maybe comparable to the way that it is insulated. This is especially relevant in the case of new and modern building designs that have already introduced advanced insulation technologies and have rather little to expect from enhancements of their shell insulation or energy systems.

The ability to measure occupancy (headcount) in a cost efficient way would add to the proliferation and validation of such concepts. This appears to be quite a serious limitation to the practical uptake and validation of operational retrofits.

In the case of occupancy, besides operational retrofits, one needs to also consider all other *ad hoc* occupancy changes, on a case per case basis, in order to see if and how they may affect the retrofit impact and should, therefore, be uptake in the adjustment methodology. For example, the change of the timing of the building use and the effective building area would also require adjustment.

4. Discussion—Further Research

The work presented here is a conceptual part of a larger ongoing research aiming at practical retrofit evaluation. The adjustment methodology presented above will be incorporated there together within another major topic: **the definition of the consumption, baseline and post retrofit, in building spaces**. Whenever our retrofit scope is that of a building space alone (a floor, a set of rooms, etc.), the measurement of the HVAC energy consumed in that space may be a very hard exercise indeed. Similar may apply if we want to look into a specific part of the retrofit strategy that has been applied to a specific part of the building. We will now need a complementary concept to manage such cases; otherwise the applicability of the adjustment methodology will be limited. In addition, the **monitoring protocol duration** is a highly important, issue. The more we can shorten it and still reach valid impact results, the more promising the methodology would be in true business environments.

Field **validation will be required for the full methodology, energy baseline and post retrofit definition and adjustment incorporation**, to monitor how the calculated impact evolves as the monitoring duration increases. We hope in this way to define an "effective duration", required for this exercise to converge a safe assessment of the impact. We also expect this to be close to that of a week.

In short, the next step of this research would be a full and **comprehensive methodology for retrofit evaluation that would be practical and short in time**, without the requirement to resort to costly and expensive measurements (e.g. velocities in air and ducts, etc.) and without the need to carry on the campaign for long periods. These two conditions would grant the methodology some significant opportunity for true industry uptake.

Acknowledgements

The work presented in this paper is partly funded by the European Commission within the 7th Framework Programme (RESSEEPE Project, Grant Agreement No: 609377).

References

[1] IPMVP, International Performance Measurement and Verification Protocol (2010) Concepts and Options for Determining Energy and Water Savings. Efficiency Valuation Organization.

[2] ASHRAE (2002) Ashrae Guideline 14: Measurement of Energy and Demand Savings. American Society of Heating, Refrigerating, and Air-Conditioning Engineers Inc., Atlanta, GA.

[3] Burkharta, M.C., Heob, Y. and Zavala, V.M. (2014) Measurement and Verification of Building Systems under Uncertain Data: A Gaussian Process Modeling Approach. *Energy and Buildings*, **75**, 189-198. http://dx.doi.org/10.1016/j.enbuild.2014.01.048

[4] Srivastava, A., Tewaria, A. and Dongb, B. (2013) Baseline Building Energy Modeling and Localized Uncertainty Quantification Using Gaussian Mixture Models. *Energy and Buildings*, **65**, 438-447. http://dx.doi.org/10.1016/j.enbuild.2013.05.037

[5] Wang, L.P., Mathew, P. and Pang, X.F. (2012) Uncertainties in Energy Consumption Introduced by Building Operations and Weather for a Medium-Size Office Building. *Energy and Buildings*, **53**, 152-158. http://dx.doi.org/10.1016/j.enbuild.2012.06.017

[6] Walter, T., Price, P.N. and Sohn, M.D. (2014) Uncertainty Estimation Improves Energy Measurement and Verification Procedures. *Applied Energy*, **130**, 230-236. http://dx.doi.org/10.1016/j.apenergy.2014.05.030

[7] Kissock, J.K. (1993) A Methodology to Measure Retrofit Energy Savings in Commercial Buildings. Ph.D. Thesis, Department of Mechanical Engineering, Texas A&M University.

[8] Krarti, M., Kreider, J., Cohen, D. and Curtiss, P. (1998) Prediction of Energy Saving for Building Retrofits Using Neural Networks. *Journal of Solar Energy Engineering*, **120**, 47-53. http://dx.doi.org/10.1115/1.2888071

[9] Dhar, A., Reddy, T. and Claridge, D. (1999) A Fourier Series Model to Predict Hourly Heating and Cooling Energy Use in Commercial Buildings with Outdoor Temperature as the Only Weather Variable. *Journal of Solar Energy Engineering*, **121**, 47-53. http://dx.doi.org/10.1115/1.2888142

[10] Reddy, T.A., Saman, N.F., Claridge, D.E., Haberl, J.S., Turner, W.D. and Chalifoux, A.T. (1997) Baselining Methodology for Facility-Level Monthly Energy Use—Part 1: Theoretical Aspects. *ASHRAE Transactions*, **103**, Part 2.

[11] Fels, M. (1986) Special Issue Devoted to Measuring Energy Savings, the Princeton Score Keeping Method (PRISM). *Energy and Buildings*, **9**, 5-18.

[12] Kaiser, M.J. and Pulsipher, A.G. (2010) Preliminary Assessment of the Louisiana Home Energy Rebate Offer Program Using IPMVP Guidelines. *Applied Energy*, **87**, 691-702. http://dx.doi.org/10.1016/j.apenergy.2009.08.001

[13] Knight, I.P. and Dunn, G.N. (2003) Evaluation of Heat Gains in UK Office Environments. Worldwide CIBSE/ASHRAE Gathering of the Building Services Industry, Edinburgh.

[14] Hoes, P., Hensen, J.L.M., Loomans, M.G.L.C., de Vries, B. and Bourgeois, D. (2009) User Behavior in Whole Building Simulation. *Energy and Buildings*, **41**, 295-302. http://dx.doi.org/10.1016/j.enbuild.2008.09.008

[15] Seligman, C., Darley, J.M. and Becker, L. (1997) Behavioral Approaches to Residential Energy Conservation. *Energy and Buildings*, **1**, 325-337. http://dx.doi.org/10.1016/0378-7788(78)90012-9

[16] Stern, P.C. (1985) Energy Efficiency in Buildings: Behavioral Issues. National Academy Press, Washington DC.

[17] Stern, P.C. (2000) New Environmental Theories: Towards a Coherent Theory of Environmentally Significant Behavior. *Journal of Social Issues*, **56**, 407-424. http://dx.doi.org/10.1111/0022-4537.00175

[18] Owens, J. and Wilhite, H. (1988) Household Energy Behavior in Nordic Countries: An Unrealized Energy Saving Potential. *Energy*, **13**, 853-859. http://dx.doi.org/10.1016/0360-5442(88)90050-3

[19] Lutzenhiser, L. (1993) Social and Behavioral Aspects of Energy Use. *Annual Review of Energy and the Environment*, **18**, 247-289. http://dx.doi.org/10.1146/annurev.eg.18.110193.001335

[20] Mullaly, C. (1998) Home Energy Use Behavior: A Necessary Component of Successful Local Government Home Energy Conservation Programs. *Energy Policy*, **26**, 1041-1052. http://dx.doi.org/10.1016/S0301-4215(98)00046-9

[21] Steg, L. (2008) Promoting Household Energy Conservation. *Energy Policy*, **36**, 4449-4453. http://dx.doi.org/10.1016/j.enpol.2008.09.027

[22] Guerra-Santin, O., Itard, L. and Visscher, H. (2009) The Effect of Occupancy and Building Characteristics on Energy Use for Space and Water Heating in Dutch Residential Stock. *Energy and Buildings*, **41**, 1223-1232. http://dx.doi.org/10.1016/j.enbuild.2009.07.002

[23] Guerra-Santin, O. and Itard, L. (2010) Occupants' Behavior: Determinants and Effects on Residential Heating Consumption. *Building Research and Information*, **38**, 318-338. http://dx.doi.org/10.1080/09613211003661074

[24] Gram-Hansen, K. (2010) Residential Heat Comfort Practices: Understanding Users. *Building Research and Information*, **38**, 175-186. http://dx.doi.org/10.1080/09613210903541527

[25] Gram-Hansen, K. (2011) Households' Energy Use—Which Is the More Important: Efficient Technologies or User Practices. *Proceedings of the World Renewable Energy Congress*, Linkoping, 8-13 May 2011.

[26] Haas, R. and Biermayr, P. (2000) The Rebound Effect for Space Heating—Empirical Evidence from Austria. *Energy Policy*, **28**, 403-410. http://dx.doi.org/10.1016/S0301-4215(00)00023-9

[27] Sunikka-Blank, M. and Galvin, R. (2012) Introducing the Prebound Effect: The Gap between Performance and Actual Energy Consumption. *Building Research and Information*, **40**, 260-273. http://dx.doi.org/10.1080/09613218.2012.690952

[28] Baker, N. and Steemers, K. (2000) Energy and Environment in Architecture: A Technical Design Guide. Taylor & Francis, London.

[29] Steemers, K. and Yun, G.Y. (2009) Household Energy Consumption: A Study of the Role of Occupants. *Building Research and Information*, **37**, 625-637. http://dx.doi.org/10.1080/09613210903186661

Permissions

The contributors of this book come from diverse backgrounds, making this book a truly international effort. This book will bring forth new frontiers with its revolutionizing research information and detailed analysis of the nascent developments around the world.

We would like to thank all the contributing authors for lending their expertise to make the book truly unique. They have played a crucial role in the development of this book. Without their invaluable contributions this book wouldn't have been possible. They have made vital efforts to compile up to date information on the varied aspects of this subject to make this book a valuable addition to the collection of many professionals and students.

This book was conceptualized with the vision of imparting up-to-date information and advanced data in this field. To ensure the same, a matchless editorial board was set up. Every individual on the board went through rigorous rounds of assessment to prove their worth. After which they invested a large part of their time researching and compiling the most relevant data for our readers.

The editorial board has been involved in producing this book since its inception. They have spent rigorous hours researching and exploring the diverse topics which have resulted in the successful publishing of this book. They have passed on their knowledge of decades through this book. To expedite this challenging task, the publisher supported the team at every step. A small team of assistant editors was also appointed to further simplify the editing procedure and attain best results for the readers.

Apart from the editorial board, the designing team has also invested a significant amount of their time in understanding the subject and creating the most relevant covers. They scrutinized every image to scout for the most suitable representation of the subject and create an appropriate cover for the book.

The publishing team has been an ardent support to the editorial, designing and production team. Their endless efforts to recruit the best for this project, has resulted in the accomplishment of this book. They are a veteran in the field of academics and their pool of knowledge is as vast as their experience in printing. Their expertise and guidance has proved useful at every step. Their uncompromising quality standards have made this book an exceptional effort. Their encouragement from time to time has been an inspiration for everyone.

The publisher and the editorial board hope that this book will prove to be a valuable piece of knowledge for researchers, students, practitioners and scholars across the globe.

List of Contributors

Ahmad Qasaimeh
Department of Civil Engineering, Jerash University, Jerash, Jordan

Mohammad Qasaimeh
Chemical Engineering Department, AlHuson University College, Al-Balqa Applied University, Salt, Jordan

Zaydoun Abu-Salem
Department of Civil Engineering, Philadelphia University, Amman, Jordan

Mohammad Momani
Department of Electrical Engineering, Yarmouk University, Irbid, Jordan

Kondakkagari Dharma Reddy, Pathi Venkataramaiah and Tupakula Reddy Lokesh
Department of Mechanical Engineering, S. V. University, Tirupati, India

V. Dlamini, R. C. Bansal and R. Naidoo
Department of Electrical Electronic & Computer Engineering, University of Pretoria, Pretoria, South Africa

Fábio Branco Vaz de Oliveira, Kengo Imakuma and Delvonei Alves de Andrade
Nuclear and Energy Research Institute, Cidade Universitária, São Paulo, Brazil

Melissa Matlock
GRID Alternatives, Riverside, USA

Esin Okay
Department of Banking and Finance, Faculty of Commercial Sciences, Istanbul Commerce University, Istanbul, Turkey

Nnenesi Kgabi
Department of Civil and Environmental Engineering, Polytechnic of Namibia, Windhoek, Namibia

Charles Grant and Johann Antoine
International Centre for Environmental and Nuclear Sciences, University of the West Indies, Kingston, Jamaica

Sumera I. Chaudhry and Manohar Das
Department of Electrical and Computer Engineering, Oakland University, Rochester, USA

Paul Ojeaga
Bergamo University, Bergamo, Italy

Deborah Odejimi
Igbinedion University, Okada, Nigeria

Emmanuel George and Dominic Azuh
Covenant University, Ota, Nigeria

Jay Zarnikau
Frontier Associates LLC, Austin, USA
LBJ School of Public Affairs and Division of Statistics and Scientific Computing, The University of Texas at Austin, Austin, USA

Shuangshuang Zhu
Frontier Associates LLC, Austin, USA

Nan Wang and Yanan Cheng
State Grid Tianjin Economic Research Institute, Tianjin, China

Wei Liang
State Grid Tianjin Electric Power Research Institute, Tianjin, China

Yunfei Mu
Key Laboratory of Smart Grid of Ministry of Education, Tianjin University, Tianjin, China

Anis Ammous
Engineering School of Sfax (ENIS), University of Sfax, Sfax, Tunisia
College of Engineering and Islamic Architecture, Umm Al Qura University, Mecca, Saudi Arabia

Hervé Morel
Université de Lyon, INSA Lyon, Lab. AMPERE, CNRS, Villeurbanne, France

Zexuan Lu and Hongsheng Xia
Management College, Jinan University, Guangzhou, China

Xi Yang and Xunmin Ou
Institute of Energy, Environment and Economy, Tsinghua University, Beijing, China
China Automotive Energy Research Center, Tsinghua University, Beijing, China

Qing Tong
Institute of Energy, Environment and Economy, Tsinghua University, Beijing, China

Talakonukula Ramesh
Government Polytechnic, Nirmal, India

Ravi Prakash
Department of Mechanical Engineering, Motilal Nehru National Institute of Technology, Allahabad, India

Karunesh Kumar Shukla
Department of Applied Mechanics, Motilal Nehru National Institute of Technology, Allahabad, India

P. D. Lezhniuk and V. A. Komar
Department of Electric Power Stations and Systems, Vinnytsia National Technical University, Vinnytsia, Ukraine

D. S. Sobchuk
Department of Electric Energy Supply, Lutsk National Technical University, Lutsk, Ukraine

A. K. Azad, M. G. Rasul, M. M. K. Khan, T. Ahasan and S. F. Ahmed
School of Engineering and Technology, Central Queensland University, Rockhampton, Australia

Podmirseg Daniel
Institute for Buildings and Energy Technical University, Graz, Austria

Amat A. Basari, Sosuke Awaji, Song Wang and Seiji Hashimoto
Division of Electronics and Informatics, Gunma University, Kiryu, Gunma, Japan

Shunji Kumagai, Kenji Suto, Hiroaki Okada, Hideki Okuno and Bunji Homma
Research and Development Department, Mitsuba Corporation, Kiryu, Gunma, Japan

Wei Jiang and Shuren Wang
School of Hydraulic, Energy and Power Engineering, Yangzhou University, Yangzhou, China

K. S. V. Swarna, G. M. Shafiullah, Amanullah M. T. Oo and Alex Stojcevski
School of Engineering, Faculty of Science, Engineering and Built Environment, Deakin University, Geelong Waurn Ponds Campus, Australia

Jeff Sommerfeld and Laurie Buys
Queensland University of Technology, Brisbane, Australia

Nikos Sakkas and Evangelos Kaltsis
Applied Industrial Technologies Ltd., Gerakas, Greece